T0185656

SpringerBriefs in Public Health

More information about this series at http://www.springer.com/series/10138

Stefania Boccia • Róza Ádány
Paolo Villari • Martina C. Cornel
Corrado De Vito • Roberta Pastorino
Editors

Personalised Health Care

Fostering Precision Medicine Advancements
for Gaining Population Health Impact

 Springer

Editors
Stefania Boccia
Section of Hygiene
University Department of Life Sciences
and Public Health
Università Cattolica del Sacro Cuore
Rome, Italy

Department of Woman and Child Health
and Public Health, Public Health Area
Fondazione Policlinico Universitario
A. Gemelli IRCCS
Rome, Italy

Paolo Villari
Department of Public Health and Infectious
Diseases
Sapienza University of Rome
Rome, Italy

Corrado De Vito
Department of Public Health and Infectious
Diseases
Sapienza University of Rome
Rome, Italy

Róza Ádány
MTA-DE Public Health Research Group
University of Debrecen
Debrecen, Hungary

Martina C. Cornel
Amsterdam University Medical Centers
location VUMC
Vrije Universiteit Amsterdam
Department of Clinical Genetics
Section Community Genetics
and Amsterdam Public Health Research
Institute
Amsterdam, The Netherlands

Roberta Pastorino
Department of Woman and Child Health
and Public Health, Public Health Area
Fondazione Policlinico Universitario
A. Gemelli IRCCS
Rome, Italy

ISSN 2192-3698   ISSN 2192-3701   (electronic)
SpringerBriefs in Public Health
ISBN 978-3-030-52398-5   ISBN 978-3-030-52399-2   (eBook)
https://doi.org/10.1007/978-3-030-52399-2

This Springer imprint is published by the registered company Springer Nature Switzerland AG
The registered company address is: Gewerbestrasse 11, 6330 Cham, Switzerland

Acknowledgements

This book is a work of the PRECeDI project (Marie Skłodowska-Curie Research and Innovation Staff Exchange—RISE No. 645740). Thanks to Dario Arzani and Rosarita Amore for editing and Laura Motta for support in the project implementation.

Contribution 2.3 was supported by the GINOP-2.3.2-15-2016-00005 project, which is co-financed by the European Union under the European Social Fund and European Regional Development Fund, as well as by the Hungarian Academy of Sciences (TK2016-78). Péter Pikó is recipient of a grant provided by the framework of the UNKP-19-3 New National Excellence Program of the Hungarian Ministry for Innovation and Technology.

List of Abbreviations

ACCE	Analytic validity, clinical validity, clinical utility, and ethical, legal, and social implications
AMSTAR	Assessment of multiple systematic reviews
AUC	Area under the curve
CDC	Centers for Disease Control and Prevention
CDSS	Clinical decision support service
CpG	Cytosine-guanine dinucleotides
CRC	Colorectal cancer
CVDs	Cardiovascular diseases
DFS	Disease-free survival
DTC-GTs	Direct-To-Consumer Genetic Tests
EGAPP	Evaluation of genomic applications in practice and prevention
EH	Essential hypertension
ELSI	Ethical, legal, and social implications
ESHG	European Society of Human Genetics
EU	European Union
EUnetHTA	European reference framework for HTA
EWAS	Epigenome-Wide Association Study
FAMCAT	Familial Hypercholesterolaemia case ascertainment tool
FDA	Food and Drug Administration
FH	Familial hypercholesterolaemia
GDPR	General data protection regulation
GP	General practitioners
GRS	Genetic risk score
GWAS	Genome-Wide Association Study
H2020	Horizon2020
HBOC	Hereditary breast and ovarian cancer
HDL-C	High-density lipoprotein cholesterol
HG	Hungarian general
HNC	Head and neck cancer
HR	Hungarian Roma

HRs	Hazard ratios
HTA	Health Technology Assessment
ICER	Institute for Clinical and Economic Review
IHC	Immunohistochemistry
LDL	Low-density lipoprotein
LEEFH	Dutch Expertise Centre for Inheritance Testing of Cardiovascular Diseases
LS	Lynch syndrome
LYG	Life years gained
miRNAs	micro RNAs
MSCA	Marie Skłodowska Curie Action
MSI	Microsatellite instability
NCDs	Non-communicable diseases
NICE	National Institute for Health and Clinical Excellence
OS	Overall survival
PH	Public health
PHG	Public health genomics
PICO	Population, intervention, comparator, outcome
PM	Personalised medicine
PRECeDI	Personalized pREvention of Chronic DIseases consortium
PRS	Polygenic risk score
QALYs	Quality-adjusted life-years
RISE	Research and innovation staff exchange
SNPs	Single nucleotide polymorphisms
T2D	Type 2 diabetes
VT	Venous thromboembolism
wGRS	Weighted GRS

Contents

About the Editors

Stefania Boccia, PhD, DSc, MSc is full professor of hygiene, preventive medicine, and public health at Università Cattolica del Sacro Cuore (UCSC) in Rome, Italy. She is the director of the Section of Hygiene of the Department of Life Science and Public Health of UCSC and president of the Public Health Epidemiology Section of European Public Health Association (EUPHA). She is a biologist by training and holds a 5-year post-graduate degree in Clinical Pathology, afterwards she obtained a Master of Science in Epidemiology and Biostatistics, and a Doctor of Science and PhD in Genetic Epidemiology at the Erasmus MC in Rotterdam, the Netherlands.

From 2016 to 2018 she was adjunct professor at Mount Sinai Medical School in New York, NY, USA.

In 2018 she founded the spin-off "Vihtali srl" (Value In Health Technology and Academy for Leadership & Innovation) at UCSC. She currently coordinates the project titled "European network staff eXchange for integrAting precision health in the health Care sysTems" (ExACT) funded by the European Commission (EC) within the H2020 Marie Sklodowska-Curie projects (MSCA-RISE). She is currently partner of several EU-funded projects; among them is "Integrating China in the International Consortium for Personalised Medicine" (IC2PerMed, CSA) as a member of the coordinating group. She is the author and coauthor of 216 scientific publications; her h-index (Scopus) is 37 and citation count is 5350.

Róza Ádány, MD, PhD, DSc is a medical doctor with specialization in the field of preventive medicine and public health. She is professor of public health at the University of Debrecen, Hungary; founding Dean of the Faculty of Public Health; leader of the Public Health Research Group funded by the Hungarian Academy of Sciences; head of the WHO Collaborating Centre on Vulnerability and Health; and president of the Public Health Genomics Section of the European Public Health Association. She also acts as president of the Hungarian Association of Public Health Training and Research Institutes, chief editor of the journal *Népegészségügy* (Public Health), editorial member of the *European Journal of Public Health*, associate editor of the *Frontiers in Public Health* journal, and vice-chair of the European

Advisory Committee on Health Research for WHO European Region. She was chairperson of the Association of Schools of Public Health in the European Region (ASPHER) between 2001 and 2003. She contributed as expert adviser to the development of the revised version of the Health for All and the draft of the Health 2020 strategies of the WHO Europe. She is the author and coauthor of 237 scientific publications; her h-index (Scopus) is 33 and citation count (Scopus) is 4634. She received several prestigious Hungarian and international awards in acknowledgement of her work.

Paolo Villari, MD, MPH is full professor of hygiene and, since 2015, is director of the Department of Public Health and Infectious Diseases at Sapienza University of Rome, Italy, where he also is a member of the Academic Senate. A graduate in Medicine and Surgery and Specialist in Hygiene and Preventive Medicine, he obtained his master's degree in Public Health at Harvard University. As former president of the Epidemiology Section of the European Public Health Association (EUPHA) in the years 2004–2006 and national delegate for Italy to the Federation of European Academies of Medicine (FEAM) from 2007 to 2013, he was Deputy Secretary General of the FEAM from 2010 to 2013. A member of the Scientific Committee at the Ministry of Health, Directorate General of Health Planning, from 2011 to 2013, he is currently Chairman of the National Committee for the Verification of measles and rubella elimination. He is responsible for international and national research projects, funded by the University, Ministry of Health, Ministry of University, EU, in the fields of epidemiology of infectious diseases, health management, and health services research. He is the author and coauthor of 165 publications (Scopus); his h-index (Scopus) is 32 and citation count (Scopus) is 3096.

Martina C. Cornel, MD, PhD is professor of community genetics and public health genomics at the Amsterdam University Medical Centers, location VUMC, in the Department of Clinical Genetics, Section of Community Genetics, and the Amsterdam Public Health Research Institute, in the Netherlands. She is a physician and epidemiologist. After 2000 she mainly worked on the responsible implementation of genetic testing and screening. She is vice-chair of the Netherlands patients' umbrella organization for rare and genetic diseases (VSOP: https://vsop.nl/). She is co-chair of the Public and Professional Policy Committee of the European Society of Human Genetics. She is a member of the Netherlands Health Council and two of its standing Committees (Population Screening and Advisory Council). She is chair of the Netherlands Program Committee Neonatal Heelprick Screening. She is editor-in-chief of the *Journal of Community Genetics*. She is the author and coauthor of 288 publications (Scopus); her h-index (Scopus) is 37 and citation count (Web of Science Core collection) is 3891.

Corrado De Vito, MD, PhD is an epidemiologist and associate professor of hygiene in the Department of Public Health and Infectious Diseases at Sapienza University of Rome, Italy. Since 2020 he is the director of the Residency of Hygiene and Preventive Medicine of Sapienza University of Rome. Corrado De Vito graduated in Medicine and Surgery and obtained a PhD in Infectious Diseases, Microbiology, and Public Health at Sapienza University of Rome. He is principal investigator or partner of several national and international projects in the fields of vaccines and vaccination, chronic care models, and public health genomics. In the latter field he is partner of the "European network staff eXchange for integrAting precision health in the health Care sysTems" (ExACT) funded by the European Commission (EC) within the H2020 Marie Sklodowska-Curie projects (MSCA-RISE). He is the author and coauthor of 100 publications (Scopus); his h-index (Scopus) is 20 and citation count (Scopus) is 1375.

Roberta Pastorino, PhD is a postdoc fellow in the Department of Woman and Child Health and Public Health—Public Health Area at Fondazione Policlinico Universitario A. Gemelli IRCCS in Rome, Italy. She is a bioengineer with a PhD in Biostatistics and a Master in Genetic Epidemiology. She is vice president of the Public Health Genomics Section of the European Public Health Association (EUPHA) and currently project manager of the project titled "European network staff eXchange for integrAting precision health in the health Care sysTems" (ExACT) funded by the European Commission (EC) within the H2020 Marie Sklodowska-Curie projects (MSCA-RISE). She is the author and coauthor of 66 scientific publications; her h-index (Scopus) is 16 and citation count is 614.

Contributors

Róza Ádány MTA-DE Public Health Research Group, University of Debrecen, Debrecen, Hungary

Stefania Boccia Section of Hygiene, University Department of Life Sciences and Public Health, Università Cattolica del Sacro Cuore, Rome, Italy

Department of Woman and Child Health and Public Health, Public Health Area, Fondazione Policlinico Universitario A. Gemelli IRCCS, Rome, Italy

Giordano Bottà Allelica S.r.l., Roma, Italia

Giovanna Elisa Calabrò Section of Hygiene, University Department of Life Sciences and Public Health, Università Cattolica del Sacro Cuore, Rome, Italy

Carolina Castagna Section of Hygiene, University Department of Life Sciences and Public Health, Università Cattolica del Sacro Cuore, Rome, Italy

Martina C. Cornel Amsterdam University Medical Centers, location VUMC, Vrije Universiteit Amsterdam, Department of Clinical Genetics, Section Community Genetics, and Amsterdam Public Health Research Institute, Amsterdam, The Netherlands

Elvira D'Andrea Division of Pharmacoepidemiology and Pharmacoeconomics, Brigham and Women's Hospital, Harvard Medical School, Boston, MA, USA

Department of Public Health and Infectious Diseases, Sapienza University of Rome, Rome, Italy

Ayşe Demirkan Department of Epidemiology, Erasmus Medical Center, Rotterdam, The Netherlands

Section of Statistical Multi-omics, University of Surrey, Guildford, UK

Department of Genetics, University Medical Center, Groningen, The Netherlands

Corrado De Vito Department of Public Health and Infectious Diseases, Sapienza University of Rome, Rome, Italy

Cornelia van Duijn Department of Epidemiology, Erasmus Medical Center, Rotterdam, The Netherlands

Nuffield Department of Population Health, University of Oxford, Oxford, UK

Carla G. van El Amsterdam University Medical Centers, location VUMC, Vrije Universiteit Amsterdam, Department of Clinical Genetics, Section Community Genetics, and Amsterdam Public Health Research Institute, Amsterdam, The Netherlands

Szilvia Fiatal Faculty of Medicine, Department of Public Health and Epidemiology, University of Debrecen, Debrecen, Hungary

Muir Gray Value Based Healthcare Programme, Department of Primary Care, University of Oxford, Oxford, UK

Elisa J. F. Houwink Department of Public Health and Primary Care (PHEG), Leiden University Medical Centre, Leiden, The Netherlands

Ilda Hoxhaj Section of Hygiene, University Department of Life Sciences and Public Health, Università Cattolica del Sacro Cuore, Rome, Italy

Anant Jani Value Based Healthcare Programme, Department of Primary Care, University of Oxford, Oxford, UK

Michael Lang McGill University Centre of Genomics and Policy, Montreal, QC, Canada

Jun Liu Department of Epidemiology, Erasmus Medical Center, Rotterdam, The Netherlands

Nuffield Department of Population Health, University of Oxford, Oxford, UK

Marco Mariani Section of Hygiene, University Department of Life Sciences and Public Health, Università Cattolica del Sacro Cuore, Rome, Italy

Roberta Pastorino Department of Woman and Child Health and Public Health, Public Health Area, Fondazione Policlinico Universitario A. Gemelli IRCCS, Rome, Italy

Péter Pikó MTA-DE Public Health Research Group, University of Debrecen, Debrecen, Hungary

Erica Pitini Department of Public Health and Infectious Diseases, Sapienza University of Rome, Rome, Italy

Tessel Rigter Amsterdam University Medical Centers, location VUMC, Vrije Universiteit Amsterdam, Department of Clinical Genetics, Section Community Genetics, and Amsterdam Public Health Research Institute, Amsterdam, The Netherlands

Jim Roldan Linkcare Health Services, Barcelona, Spain

Annalisa Rosso Department of Public Health and Infectious Diseases, Sapienza University of Rome, Rome, Italy

Michele Sassano Section of Hygiene, University Department of Life Sciences and Public Health, Università Cattolica del Sacro Cuore, Rome, Italy

Brigid Unim Department of Public Health and Infectious Diseases, Sapienza University of Rome, Rome, Italy

Paolo Villari Department of Public Health and Infectious Diseases, Sapienza University of Rome, Rome, Italy

Ma'n H. Zawati McGill University Centre of Genomics and Policy, Montreal, QC, Canada

Chapter 1
Introductions

Stefania Boccia

Personalization of healthcare is a driver of innovation in research, healthcare systems and industry. While approaches of personalized medicine (PM) are already being implemented especially in the fields of disease diagnosis and treatment with the use of biomarkers, development and implementation of such approaches for **disease prevention** are still at infancy. This leads to the concept of **precision health** (Chowkwanyun et al. 2018), which includes PM but broadly expands its focus to healthy individuals.

According to the latest report on the Health Status of Europe, the main causes of death in European Union (EU) countries are represented by chronic diseases, including circulatory diseases and various types of cancer, followed by respiratory diseases (Health at a Glance: Europe 2018). Additionally, the report estimated that 790, 000 people in EU countries died prematurely in 2016 due to tobacco smoking, harmful consumption of alcohol, unhealthy diets and lack of physical activity. As highlighted by the European Steering Group on Sustainable Healthcare (Acting Together: A Roadmap for Sustainable Healthcare 2016), the implementation of **sustainable healthcare** requires a shift from treatment of established diseases, to **diseases prevention and early diagnosis**, and it relies on the need to **engage citizens** to take greater responsibility for their health. In fact, despite the increase in life expectancy in the last 50 years, the latest data indicate that the average number of years of life lived with some disability in EU is 18 (Eurostat). This situation brings

S. Boccia (✉)
Section of Hygiene, University Department of Life Sciences and Public Health, Università Cattolica del Sacro Cuore, Rome, Italy

Department of Woman and Child Health and Public Health, Public Health Area, Fondazione Policlinico Universitario A. Gemelli IRCCS, Rome, Italy
e-mail: stefania.boccia@unicatt.it; stefania.boccia@policlinicogemelli.it

S. Boccia et al. (eds.), *Personalised Health Care*, SpringerBriefs in Public Health, https://doi.org/10.1007/978-3-030-52399-2_1

a huge economic burden to EU healthcare systems, thus challenging their sustainability and forcing policymakers to identify new models of disease prevention and healthcare assistance. The relevance of investing more in protection and prevention has been also addressed in the latest Health Status of Europe (Health at a Glance: Europe 2018), which further pointed out at the needs of continuing efforts in addressing health inequalities. In fact, people with a low level of education expect to live 6 years less than those with a high level of education, and the new technologies available including those discussed in this book, should not becoming another reason for which vulnerable people are left behind.

Given the potential for preventive efforts in postponing the onset of disabilities and reducing premature mortality at a reasonable cost, the expectation is that the current "one-size-fits-all" approaches take advantage of the new technologies in healthcare in order to be targeted at those who need more (Ricciardi and Boccia 2017). The expectations to realize such a personalized approach in preventive healthcare are not new. Already in 2008 an editorial reported that "*if preventive care could be provided only to those who are going to get the illness (more precise we would say now), it would be more effective and cost-effective*" (Cohen et al. 2008). More than 10 years later, however, the implementation of personalized approaches in preventive healthcare is still lagging behind.

Laying the Foundation for Making Personalized Prevention a Reality. In the current book, we present the main results of The Personalized pREvention of Chronic DIseases consortium (PRECeDI) project, and its implication in view of the most recent advancement of the research efforts in the field of personalized prevention. PRECeDI is a Marie Skłodowska Curie Action (MSCA) project funded within the Research and Innovation Staff Exchange (RISE) scheme that aimed at providing high-quality, multidisciplinary knowledge through training and research in PM, with specific reference to personalized prevention of chronic diseases (Boccia et al. 2019). The PRECeDI consortium consists of 8 beneficiaries and 3 partners, of which 7 are academic institutions and 4 non-academic, including 2 SMEs. The partners from Europe are: Institute of Public Health, Università del Sacro Cuore, Rome, Italy; Better Value Healthcare Ltd., Oxford, United Kingdom; Department of Infectious Diseases and Hygiene, Università La Sapienza, Rome, Italy; Section Community Genetics, VU University Medical Center Amsterdam, The Netherlands; LINKCARE Health Services S.L., Barcelona, Spain; Erasmus Universiteit Medisch Centrum, Department of Epidemiology, Rotterdam, The Netherlands; Public Health Research Institute, Debreceni Egyetem, Debrecen Hungary; European Public Health Association, Utrecht, The Netherlands; Myriad Genetics Srl, Milan, Italy. Outside of Europe partners included are Icahn School of Medicine at Mount Sinai, New York, U.S.A.; and Centre of Genomics and Policy Department of Human Genetics, McGill University, Montreal, Canada.

The PRECeDI consortium received funding from the Horizon2020 (H2020) EU's Eight Framework Programme for research. During 4 years (2014–2018), 28 early stage researchers and 22 experienced researchers were seconded across partners' institutions, for training in research projects related to personalized prevention of chronic diseases including cancer, cardiovascular diseases (CVDs) and

Alzheimer's disease. Different projects were implemented, from basic research to economic evaluations, from health service organization issues to physician education, including ethical, social and policy issues in PM, supported by a team of leading EU scientists.

How PRECeDI contributes to the integration of Personalized Medicine in the Prevention of Chronic Diseases. Based on the results of the research carried out by the PRECeDI consortium, a set of recommendations for policymakers, scientists and industry has been drawn up and reported in Fig. 1.1, with the main goal to foster the integration of PM approaches in the field of chronic disease prevention (Boccia et al. 2019) fig. 1.1. The **recommendations fall within the five research domains of PRECeDI,** which are:

PRECeDI Domains	PRECeDI Recommendations
 **Domain 1:** Identification of biomarkers for the prevention of chronic disease.	**R1.** Personalized interventions for the prevention of chronic diseases require robust evidence of efficacy and/or effectiveness of the new technology when implemented in health care.
 **Domain 2:** Economic evaluation of predictive genomic applications.	**R2.** In addition to what reported in R1, a comprehensive evaluation of the value (outcomes/cost) of the new technology should also include evidence on the social aspects, and context-related dimensions to better support the clinical decision-making process. Genetic or genomic applications with evidence of efficacy, effectiveness and cost-effectiveness should be implemented in clinical and public health practice.
 **Domain 3:** Ethico-legal and policy issues surrounding personalized medicine.	**R3.** The era of genomics requires that we clarify and validate the obligations and responsibilities of the research community, research participants, and the general public including patients through collaboration and dissemination of high-quality ethical, policy and legal analysis.
 **Domain 4:** Sociotechnical analysis of the pros-and cons of informing healthy individuals on their genome.	**R4.** A dedicated effort is necessary to stimulate further implementation of evidence-based interventions in health care, such as testing of family members in cases of hereditary cancers or cardiovascular diseases.
 **Domain 5:** Identification of organizational models for the provision of predictive genomic applications.	**R5.** The integration of genomic sciences in other medical specialties should be promoted through new delivery models involving different healthcare professionals and new professional roles, in order to guarantee the use and sustainability of existing and new genomic applications in practice.

Fig. 1.1 Recommendations from the PRECeDI project

1. Identification of biomarkers for the prevention of chronic diseases
2. Economic evaluation of predictive genomic applications
3. Ethico-legal and policy issues surrounding PM
4. Sociotechnical analysis of the pros and cons of informing healthy individuals on their genome
5. Identification of organizational models for the provision of predictive genetic testing.

In the present book we report, although in a non-exhaustive fashion, the main results of the research projects carried out in the five research domain that informed each of the five recommendations. The Box 1.1 reports additional details within each of the five recommendations relased.

Box 1.1: Recommendations List

Recommendation (R) 1 is based on the "Identification of biomarkers for the prevention of chronic diseases" research domain.

Biomarkers have the potential to stratify populations because they can help to indicate an individual's risk or resistance to disease as well as the potential response the individual may have to different treatments. There is also an expectation that this may lead to better targeting of preventive interventions by defining the disease and targeting the treatment based on a person's molecular pathology.

R1. Personalized interventions for the prevention of chronic diseases require robust evidence of efficacy and/or effectiveness of the new technology when implemented in healthcare.

In particular: large trials evaluating the efficacy of disease risk communication based on broad range newly discovered biomarkers (versus risk communication based on the solely traditional risk factors) on behavioural change among healthy subjects at increased risk are required for targeted evidence-based primary preventive interventions. For biomarkers that allow discriminating high-risk subjects, large trials evaluating the efficacy of medical interventions are required among such high-risk subjects for targeted evidence-based primary and secondary preventive interventions.

Where intervention studies cannot be performed, however, the use of large datasets, Big Data from collaborative research projects, should be considered for the evidence of effectiveness. In order to ensure timely results for the use of such predictive biomarkers, the collection of such evidence by action research should be foreseen in the course of implementation and accompanied by collection of genetic data to allow for state-of-the Mendelian Randomization studies to mimic conventional trials.

In these situations, a clear commitment to hypothesis to be tested in advance is needed as is the case with (the registration of) classical trials.

For tertiary prevention, the adoption of accurate biomarkers for precise monitoring and early prediction of disease progression should be encouraged.

Recommendation 2 is based on the "Economic evaluation of predictive genomic applications" research domain.

The growing availability of genomic technologies is contributing to the shift of the medical approach towards personalized medicine, where medical decisions are based on an individual's characteristics, including the genomic profile. This has made the assessment of the performance of genomic tests crucial for clinical and public health practice. In fact, in order to maximize population health benefits, it is essential to distinguish genomic tests with proven efficacy and/or effectiveness and cost- effectiveness and support their implementation.

R2. A comprehensive evaluation of the value (outcomes/cost) of genetic and genomic applications should include evidence on the efficacy and/ or effectiveness of the new technology (i.e. analytic validity, clinical validity, clinical utility), social aspects (ethical, legal and social implications, and personal utility), and context-related dimensions (e.g. economic evaluation, delivery models, organizational aspects and consumer viewpoint) to better support the decision-making process.

Genetic or genomic applications with evidence of efficacy, effectiveness and cost-effectiveness should be implemented in clinical and public health practice (i.e. programmes that include tools for identifying affected women at higher risk for inherited breast and ovarian cancers or familial history-based screening for *BRCA1/2;* universal or < 70 years of age-targeted colorectal cancer-based Lynch Syndrome screening; cascade screening of familial hypercholesterolaemia). The genomic or genetic testing programmes and their implementation should be developed and pursued based on the characteristics of target populations and healthcare systems to ensure an appropriate translation of evidence into the "real-world".

The implementation of a genetic or genomic application should be continuously assessed, measuring the population health impact and relative value of new technologies.

Adherence to the programmes should be monitored and the education and training of clinical and public health professionals should be promoted with the aim of reducing inappropriate use in healthcare.

Recommendation 3 is based on the "Ethico-legal and policy issues surrounding personalized medicine" research domain.

There is an increasing need for a coordinated effort to foster the development and further harmonization of dedicated policies to integrate genomics policies into existing health systems in a responsible manner. Introducing a common ethically and legally validated policy framework could represent one of the drivers needed to manage a future with increasingly personalized healthcare and a shift in the use of genomic approaches from disease treatment to prevention.

R3. The era of genomics requires that we clarify and validate the obligations and responsibilities of the research community, research participants and the general public.

This can be achieved through collaboration and dissemination of high-quality ethical, policy and legal analysis. Legal interoperability is necessary to ensure complementarity of goals between researchers in different jurisdictions.

In order to be at the forefront of the currently shifting research landscape, we need to draw on multiple levels of expertise (e.g. law, ethics, medicine, bioinformatics, IT) in an array of multidisciplinary, jurisdictional and institutional settings.

Finally, a metric assessing the impact of policy development or lack thereof is a fundamental tool to fine-tune guidance to multiple stakeholders.

Recommendation 4 is based on the "Sociotechnical analysis of the pros and cons of informing healthy individuals on their genome" research domain.

Genetic testing of family members of patients affected with hereditary cancers or CVDs allows for personalized prevention and it is paramount to find and inform these family members in a timely manner. Several countries are building cascade screening programmes and they are discussing how family members can be traced and informed in an ethically responsible and efficient manner. In conditions where genetic testing offers a substantial and quantifiable risk estimate and prevention is available, preventive services should be prioritized. More government involvement is needed as a formally organized screening programme could standardize support and information, and lead to more equitable healthcare.

R4. A dedicated effort is necessary to stimulate further ethically responsible implementation of evidence-based interventions in healthcare, such as testing of family members in cases of hereditary cancers or cardiovascular diseases.

Where guidelines for such genetic testing exist, collaboration between genetic and non-genetic healthcare professionals needs to be facilitated to improve implementation, education opportunities must be provided and roles and responsibilities towards informing family members must be reconsidered so we can achieve a truly multidisciplinary approach that can realize the potential of personalized medicine.

Recommendation 5 is based on the "Identification of organizational models for the provision of predictive genetic testing" research domain.
The identification and evaluation of existing genetic service delivery models are important steps towards the enhancement and standardization of genetic service provision. Integration of genetics in all medical specialties, collaboration among different healthcare professionals, and redistribution of professional roles are fundamental elements for the organization of these models. Furthermore, their implementation must hinge on professional education, adequate funding and public awareness in the field of genomic medicine.

R5. The integration of genetics in other medical specialties should be promoted through new delivery models involving different healthcare professionals (medical specialists, nurses, technicians, etc.) and new professional roles (i.e. genetic counsellors, genetic associates, genetic nurses), in order to guarantee the use and sustainability of existing and new genomic applications in practice.

Roles and responsibilities (e.g. risk assessment, genetic counselling, genetic testing) should be redistributed among different health professionals to enhance work performance and the standard of care.

It is advisable to define the appropriate model for genetic service provision in a specific setting according to the type of healthcare system and the genetic test provided.

Professional education/training in genomics medicine, laboratory quality standards and public awareness are essential factors for the successful implementation of genomic applications in practice.

Taking into account that personalized prevention can only be successfully implemented when handled as a truly cross-sectoral topic, our recommendations integrate the perspective of experts across the entire healthcare value chain that are represented in the PRECeDI consortium. The belief is that the implementation of the recommendations will benefit citizens, patients, healthcare professionals, healthcare authorities and industry, and ultimately will seek to contribute to better health for Europe's citizens (Boccia and Ricciardi 2019).

References

Acting Together: A Roadmap for Sustainable Healthcare. (2016). Retrieved from https://www.sustainable-healthcare.com/content-assets/uploads/2016/10/Sustainable-Healthcare-Brochure-_-Dec-2016-_-For-download_ESG-Section.pdf. Accessed 23 Dec 2019.

Boccia, S., & Ricciardi, W. (2019). Personalized prevention and population health impact: how can public health professionals be more engaged?. *European Journal of Public Health, 30*(3), 391–392.

Boccia, S., Pastorino, R., Ricciardi, W., et al. (2019). How to integrate personalized medicine into prevention? Recommendations from the Personalized Prevention of Chronic Diseases (PRECeDI) Consortium. *Public Health Genomics, 22*(5-6), 208–214.

Chowkwanyun, M., Bayer, R., & Galea, S. (2018). "Precision" public health - between novelty and hype. *The New England Journal of Medicine, 379*(15), 1398–1400.

Cohen, J. T., Neumann, P. J., & Weinstein, M. C. (2008). Does preventive care save money? Health economics and the presidential candidates. *The New England Journal of Medicine, 358*(7), 661–663.

Eurostat. (n.d.). Retrieved from https://ec.europa.eu/eurostat/web/health/data. Accessed 23 Nov 2019.

Health at a Glance: Europe 2018 State of Health in the EU. (2018). Retrieved from https://ec.europa.eu/health/sites/health/files/state/docs/2018_healthatglance_rep_en.pdf. Accessed 16 Jan 2020.

Ricciardi, W., & Boccia, S. (2017). New challenges of public health: Bringing the future of personalized healthcare into focus. *European Journal of Public Health, 27*(Suppl_4), 36–39.

Chapter 2
Identification of Biomarkers for the Prevention of Chronic Disease

Stefania Boccia, Jun Liu, Ayşe Demirkan, Cornelia van Duijn, Marco Mariani, Carolina Castagna, Roberta Pastorino, Szilvia Fiatal, Péter Pikó, Róza Ádány, and Giordano Bottà

2.1 Identification of Novel Biomarkers for Primary and Secondary Prevention of Diabetes in the –Omics Era

Jun Liu, Ayşe Demirkan, and Cornelia van Duijn

Early lifestyle intervention is a cost-effective recommendation to reduce the incidence of type 2 diabetes (T2D), asking for informative, sensitive and specific markers. Although the standard laboratory tests, such as fasting glucose, 2-h post-prandial glucose and glycated haemoglobin A1c (HbA1c), provide strong evidence for the risk of T2D, these predictors emerge after years of subclinical metabolic dysfunction that may induce atherosclerosis but also pathology to key organs such as the kidney, liver, brain and eye. Traditional risk factors such as age, sex, body mass index and waist circumference also explain considerable part of future risk but fail to capture the full complexity of the aetiology and their predictive performance varies between different risk groups (Kengne et al. 2014). There is an increasing

S. Boccia
Section of Hygiene, University Department of Life Sciences and Public Health, Università Cattolica del Sacro Cuore, Rome, Italy

Department of Woman and Child Health and Public Health, Public Health Area, Fondazione Policlinico Universitario A. Gemelli IRCCS, Rome, Italy
e-mail: stefania.boccia@unicatt.it; stefania.boccia@policlinicogemelli.it

J. Liu · C. van Duijn
Department of Epidemiology, Erasmus Medical Center, Rotterdam, The Netherlands

Nuffield Department of Population Health, University of Oxford, Oxford, UK
e-mail: jun.liu@ndph.ox.ac.uk; cornelia.vanduijn@ndph.ox.ac.uk

© The Editor(s) and Author(s), under exclusive license to Springer Nature Switzerland AG 2021
S. Boccia et al. (eds.), *Personalised Health Care*, SpringerBriefs in Public Health, https://doi.org/10.1007/978-3-030-52399-2_2

interest in finding informative markers that indicate the particular metabolic dysfunctions before the manifestation of the disease. This would open avenues to identify people at high risk and enable new preventive lifestyle interventions or early treatments targeted to their individual molecular profile, personalising prevention and treatment.

In the past decade, high-throughput technology has fuelled the research on biomarkers of T2D. Large-scale research on the molecular features collected in epidemiological and clinical biobanks is possible and has been extremely successful. The collection of research tools referred to as *omics technology* involves the comprehensive characterisation, quantitation and quantification of biological molecules. It includes but is not limited to genomics including also epigenomics, transcriptomics, proteomics, metabolomics and microbiomics. Exploring the omics data has accelerated our insights into biological pathways relevant to the aetiology and pathological progression of T2D, hence improving the prevention of T2D.

This section introduces several omics approaches, focusing on genomics, transcriptomics, epigenomics and metabolomics, to study pathophysiology and mechanism of T2D for better prevention of T2D.

A. Demirkan
Department of Epidemiology, Erasmus Medical Center, Rotterdam, The Netherlands

Section of Statistical Multi-omics, University of Surrey, Guildford, UK

Department of Genetics, University Medical Center, Groningen, The Netherlands
e-mail: a.demirkan@surrey.ac.uk

M. Mariani (✉) · C. Castagna
Section of Hygiene, University Department of Life Sciences and Public Health,
Università Cattolica del Sacro Cuore, Rome, Italy
e-mail: marco.mariani02@icatt.it; carolina.castagna01@icatt.it

R. Pastorino
Department of Woman and Child Health and Public Health, Public Health Area, Fondazione
Policlinico Universitario A. Gemelli IRCCS, Rome, Italy
e-mail: roberta.pastorino@policlinicogemelli.it

S. Fiatal
Faculty of Medicine, Department of Public Health and Epidemiology, University of
Debrecen, Debrecen, Hungary
e-mail: fiatal.szilvia@med.unideb.hu

P. Pikó · R. Ádány
MTA-DE Public Health Research Group, University of Debrecen, Debrecen, Hungary
e-mail: piko.peter@med.unideb.hu; adany.roza@med.unideb.hu

G. Bottà
Allelica S.r.l., Roma, Italy
e-mail: giordano@allelica.com

Genomics

Genomics has been one of the first disciplines that have been applied successfully at high throughput in the research of T2D and other disorders. Genome-wide association study (GWAS) is one of the most successful approaches in genomic research of complex disorders such as T2D. GWAS is a hypothesis-free observational method exploring the association of genetic variants to a target trait. The basic rationale of GWAS is that if a variant is causally associated with a disease, the variant is expected to be found more often in cases than controls. The first successful GWAS analyses were conducted as part of the Wellcome Trust Case Control Consortium (WTCCC), which was established to harness the power of newly available genotyping technology (WTCCC 2007). The study included 2000 T2D cases and 3000 nationally ascertained controls with up to 500,000 sites of genome sequence variation (single-nucleotide polymorphisms (SNPs)) and detected three T2D loci: *PPARG, KCNJ11* and *TCF7L2* (Wellcome Trust Case Control Consortium 2007).

The size of the GWAS studies has grown rapidly over the year, accelerating the numbers of loci identified. Using the large-scale and accessible resources, such as the UK Biobank, and the dense and accurate genotype imputation, e.g. 1000 Genomes and Haplotype Reference Consortium (HRC), the number of genetic determinants of T2D identified through GWAS has grown from three to 243 (Mahajan et al. 2018). With a large number of genetic variants identified, the genome-wide SNP heritability explains up to 18% of the risk of T2D, which accounts for approximately half the median estimates of heritability derived from twin and family studies (Mahajan et al. 2018). The genetic variants can discriminate and predict the population with a high risk of T2D: the individuals in the top 2.5% polygenic risk score have a 9.4-fold increased risk of T2D compared with the individuals in the bottom 2.5% (Mahajan et al. 2018). In that, genomic research has achieved one of the goals of the complex genetic research: we can now stratify the population and identify people whose risks of diabetes is increased due to a combination of genes to a very similar extend as that of persons who carry major Mendelian mutations in genes such as Low Density Lipoprotein (LDL) receptor. As these major mutations justify PM, e.g. cascade screening based on family history and starting lipid treatment of carriers in early adolescence, one of the goals of the population health research will be to explore the opportunities of early interventions in this subgroup with a similar increase in risk. There are many questions that remain to be answered: 1) we often do not know the causal variant but rather a region in which the causal variant is located; 2) we will have to determine what is the onset age of these persons; 3) we will have to determine what is the cause of the disease: whereas the mutations in the LDL receptor gene point to a single pathway, the combination of genetic factors in those of the extreme of the polygenic risk scores (PRSs) confront us with the complexity and complex disease. We need to explore multiple pathways which only can be done by new technologies such as transcriptomics, proteomics and metabolomics.

Transcriptomics

One of the most important lessons of GWAS decade is that the genetic variants implicated in T2D and other complex disorders are for a large part found in regulatory regions of the genome as is the case for the genetic variants implicated in obesity (Mahajan et al. 2018; Wahl et al. 2017). Regulation of gene expression normally happens at the level of RNA biosynthesis (transcription) and is accomplished through the sequence-specific binding of proteins (transcription factors) that activate or inhibit transcription. A variety of biological molecules may bind to the RNA to alter the regulation, including proteins (e.g. translational repressors and splicing factors), other RNA molecules (e.g. micro-RNAs (miRNAs) and small molecules. GWAS has been integrating transcriptomic research to fine-map causal variants based on conditional analysis and, perhaps more importantly, by combining the GWAS with transcriptomic data and data on protein function. Using the power of very large transcriptomic data sets substantially improved the fine-mapping of causal variants of T2D in the set of 243 genetic variants identified to date (Mahajan et al. 2018), resulting in 51 signals. The study also succeeded in identifying eighteen genes (19 coding variants) were validated as the therapeutic targets (Mahajan et al. 2018). These cross-omics analyses provided novel insights into the biological mechanisms of T2D and identified genes and proteins operating at the fine-mapped regulatory signals. The establishment of causal variants and variant-gene links through genomics studies provide targets to modify in cellular and animal in vivo models. Thus, the clinical translation of the genomic associations is progressing. However, a major problem has been that genetic regulation of transcription for a given messenger RNA may be tissue specific. The paucity of tissue that may be relevant for the pathogenesis of diabetes, such the liver, pancreas, and muscle, is limited. Single-cell sequencing is allowing us to fill these gaps and an important development is the human cell atlas aiming to create comprehensive reference maps of all human cells as a basis for both understanding human health and diagnosing, monitoring, and treating disease (https://www.humancellatlas.org/) using state-of-the-art –omics research. This atlas may fill important physiological gaps in our knowledge and fuel population health research in the next decade.

Epigenomics

An important driver of the genetic regulation that is of great interest to epidemiological and clinical research of T2D is epigenetics. Epigenetic changes involve histone modification, non-coding RNA and DNA methylation. Epigenetic changes can modify the RNA expression or transcription factor binding and may relate to the risk and progression of the disease. Most epidemiological studies focus on DNA methylation at cytosine-guanine dinucleotides (CpG) sites, which can be captured high-throughput by arrays. Similar to GWAS, epigenome-wide association study

(EWAS) detect differences in DNA methylation in a cost-efficient, high-throughput and accurate way, in relation to the target disease or phenotype. Epigenetic modifications may occur in early phases of the pathology of T2D: DNA methylation has been associated with T2D but also with fasting glucose and insulin, and the methylation risk score of T2D has predicted incident cases beyond traditional risk factors including obesity and waist-hip ratio (Wahl et al. 2017). A recent longitudinal study published in *Nature Medicine* reported that most DNA methylation changes occur 80–90 days before detectable glucose elevation (Chen et al. 2018). This suggests that differential DNA methylation may evoke changes in glucose and is involved in the early stage(s) of diabetes.

A number of blood-based EWAS of T2D and related traits, such as fasting glucose and insulin, were published. But only a limited number of methylation sites have been discovered and replicated for T2D and related outcomes, including the sites in *TXNIP, ABCG1, CPT1A, PHOSPHO1, ABCG1* and *SREBF1. TXNIP* is the replicated methylation loci associated with T2D and the role in T2D has been studied in detail by functional studies. Differential methylation at CpG in *CPT1A* and *ABCG1* are also associated with T2D or related traits such as circulating lipids and obesity. When studying the effect of epigenetic in glucose metabolism, we identified CpGs in *LETM1, RBM20, IRS2, MAN2A2* and the 1q25.3 region associated with fasting insulin, and in *FCRL6, SLAMF1, APOBEC3H* and the 15q26.1 region with fasting glucose (Liu et al. 2019). Differential methylation explains at least 16.9% of the association between obesity and insulin (Liu et al. 2019). We integrated the epigenomic findings with that of genomics and transcriptomics and found evidences that differential methylation is in the crosstalk between the adaptive immune system and glucose homeostasis. Findings are promising but epigenetic research is still in an early phase, asking for large-scale efforts to fill the gap that are ongoing such as the epigenomics road map (http://www.roadmapepigenomics.org/) and the diabetes epigenome atlas (https://www.diabetesepigenome.org/).

Metabolomics

Metabolomics is the systematic study of chemical compounds (referring to as *metabolites* below) that are the result of cellular processes. Metabolomics aims to quantify the concentration of metabolites, the small molecule intermediates and products of metabolism in various tissues. The high-throughput techniques including mass spectrometry and nuclear magnetic resonance offer the opportunity to determine various chemical molecules at large scale and at a low-cost price. As a typical metabolic disorder, T2D disturbs the internal homeostasis of the organism, which may result in early-detectable alteration (Tabák et al. 2009). Understanding the organisation of metabolome and T2D may help to discover biomarkers for the risk and consequences of T2D. As is the case for epigenetics, metabolic changes related to the disease onset or progression may be tissue specific.

One of the major classes of metabolites that have been studied in relation to T2D is small molecules such as amino acids and lipids. Circulating branched amino acids have been implicated in T2D and related by various epidemiological studies. Wang et al. reported in *Nature Medicine* in 2011 (Wang et al. 2011) that five amino acids, i.e. leucine, isoleucine, valine, tyrosine and phenylalanine, are important predictors of the occurrence of diabetes with the follow-up of 2422 non-diabetic individuals for 12 years (Wang et al. 2011). Other metabolites, such as glutamate, glycine, 2-aminoadipic acid or α-hydroxybutyrate, serum saturated fatty acids, glycerolipids, sphingolipids and phospholipids have been found to be strongly related to T2D as well. An important question from an epidemiologic perspective is to what extent these metabolites improve the prediction of T2D or are associated with its progression. It has been shown that with T2D associated metabolites, the prediction of T2D was improved significantly. One of our previous studies showed that the combined effect of 24 metabolites including ten lipoprotein sub-fractions yield a powerful discrimination model for predicting future T2D. The combined metabolite model predicts future T2D better than fasting glucose in the population who are female, younger than 50 years, or those with normal weight (Liu et al. 2017).

Conclusions

In summary, as the high-throughput era in omics is progressing, genomics and other omics will be effective in disentangling the aetiology and progression of T2D, yielding insights of the pathophysiology, identifying more biomarkers for the prevention and treatment of T2D. As a consequence, there is urgency to integrate – omics research in population health.

2.2 The Prognostic Role of Micro-RNAs in Head and Neck Cancers: An Umbrella Review

Marco Mariani, Carolina Castagna, Roberta Pastorino, and Stefania Boccia

Head and neck cancer (HNC) is a heterogeneous group of neoplasms which develops from different tissues such as oral and nasal cavities, paranasal sinuses, pharynx, and larynx. It represents the sixth most common cancer and the seventh cause of cancer-related deaths worldwide (Kumarasamy et al. 2019). The complex anatomy of this region results in complicated patterns of tumour invasion and consequently in difficulties in treating patients suffering from these diseases (Chin et al. 2006). The majority of patients present already advanced stages of cancer at diagnosis, characterised by local aggressiveness and high potential for local and systemic metastasis (Babu et al. 2011).

Because of HNC high mortality and morbidity, a support from the development of new biomarkers and personalised care for patients is needed (Kumarasamy et al. 2019).

The role of micro-RNAs (miRNAs), as new epigenetic biomarkers, aimed at improving early diagnosis, predicting prognosis and establishing effective cancer therapies, has recently received considerable attention (Thomas et al. 2005). miR-NAs represent a class of highly conserved non-coding small RNAs that could regulate gene expression: clinical studies highlight that many miRNAs were upregulated or downregulated in a different variety of cancers and are involved in several biological processes, such as cellular proliferation, differentiation, migration, apoptosis, survival and morphogenesis (Carleton et al. 2007).

Several original studies, systematic reviews and meta-analysis have been conducted in relation to miRNAs as biomarkers for cancer prognosis in several tumours such as HNC. Even if the body of evidence in relation to miRNAs and HNC prognosis is growing, there is wide variability of study aims, measurement methods, tumour sites and miRNAs among studies (Kumarasamy et al. 2019).

Hence, the aim of this study was to conduct an umbrella review, i.e. a review of reviews that compiles all the evidence from existing reviews, to investigate on the prognostic role of miRNAs as biomarkers in the field of tertiary prevention of HNC, in order to give a high-level overview.

Methods

We conducted this umbrella review synthesising the findings of systematic reviews already available in literature, retrieved on electronic databases (Pubmed, ISI Web of Science, and Scopus) from their inception to December 2019. We used the Population, Intervention, Comparator, Outcome (PICO) model search strategy and used the following terms to build the search string: systematic review, meta-analysis, cwancer, miRNAs, prognosis, survival, recurrence and relapse. Two researchers (C.C. and M.M.) independently screened the titles and abstracts and selected articles for full text review. Articles were eligible if they were systematic review with meta-analysis of observational studies that reported quantitative prognostic measures, Hazard Ratios (HRs), Overall Survival (OS) or Disease-Free Survival (DFS). The methodological quality of the included reviews was assessed using the Assessment of Multiple Systematic Reviews (AMSTAR) 2 tool (Shea et al. 2017), which consists of 16 domains presented in the form of questions. The possible answers are "Yes" if it denotes a positive result; the article presents a weakness if the answerer is negative "No" (or it cannot be provided) or "Partial Yes" in case of partial adherence to the standard. The AMSTAR overall judgement was based on the assessment of specific critical domains as: presence of a protocol registered before the commencement of the review, evaluation of the risk of bias of the studies included, appropriateness of meta-analytic methods if applicable, consideration of risk of bias when interpreting the results of the review, assessment of presence and likely impact of publication bias. The final quality judgement (high, moderate, low, critically low) was performed by two researchers and disagreements were overcome by consensus. The following data were extracted from the studies: name of the first author, publication year, country, number of qualitative studies and number of patients included in the review, miRNAs studied, level of regulation of miRNAs, primary study OS and DFS reported in the qualitative synthesis, OS (HR) and DFS (HR) of the quantitative synthesis. Results reporting a p-value < 0.05 were considered statistically significant.

Results

The search yielded 662 studies, 351 left after duplicate removal, eventually 6 systematic reviews were included (Table 2.1). According to the qualitative assessment tool AMSTAR2, 1 out of 6 systematic reviews (16.7%) was evaluated as "critically low" (Lubov et al. 2017) because it presented more than one weakness among the critical domains previously described; 4 of them (66.6%) were evaluated as "low" (Fu et al. 2011; Jamali et al. 2015; Sabarimurugan et al. 2018; Troiano et al. 2018) since they presented only one critical flaw; 1 review (16.7%) resulted as "moderate" because it had more than one weakness but no critical flaws. (Kumarasamy et al. 2019).

The total population among included studies varied from 422 to 6834 individuals. The most reported miRNAs were: miR21 from all the included studies; the Let7 family (c, d and g), miR17, 18 family (a and b), 20a, 29 family (a, b, c), 125b, 375, 451 by 3 (50%) reviews; eventually miR34a, 155, 181, 205, 210, 218, 363 by 2 (33%) reviews, the others reported by a single review.

Two studies (Kumarasamy et al. 2019; Sabarimurugan et al. 2018) performed a meta-analysis, pooling HRs (95% CIs) from different miRNAs to get an overall estimate of the effect of the combination of more miRNAs. One (Sabarimurugan et al. 2018) reported that OS from upregulated miRNAs (from 25 primary studies) have a pooled HR of 1.76 (95% CI 1.43–2.17); downregulated ones (from 20 primary studies) have a pooled HR of 2.02 (95% CI 1.43–2.17). The other (Kumarasamy et al. 2019) reports OS, taking together up- and downregulated, miRNAs have an overall HR of 1.20 (95% CI: 0.89–1.60), and when stratified by level of regulation, upregulated miRNAs have a pooled HR 4.64 (95% CI, 1.05–2.58); downregulated ones HR 0.94 (95% CI 0.65–1.39). For what concerns the DFS, in this latter study (Kumarasamy et al. 2019), when up- and downregulated miRNAs were taken together, they showed an overall HR of 2.60 (95% CI: 1.91–3.51), in particular upregulated miRNAs only (from 10 studies) showed a pooled HR 2.64 (95% CI 1.93–3.62), downregulated ones HR 2.10 (95% CI 0.71–6.20). A total of 4 reviews (Fu et al. 2011; Lubov et al. 2017; Jamali et al. 2015; Kumarasamy et al. 2019) assessed miR-21 expression in HNC patients, all showing its upregulation. The pooled HRs from the OS analysis ranged from 1.46 to 1.81 and all were statistically significant. Among these four, two (Jamali et al. 2015; Kumarasamy et al. 2019)

Table 2.1 Characteristics and main findings of the six systematic reviews and meta-analysis included

Author	Year	Number of qualitative studies and number of patients	miR	Level of regulation	Qualitative studies		Quantitative studies	
					OS	DFS	OS HR (95% CI)	DFS HR (95% CI)
Fu X	2011	4, 422	21	U	–*		1.46 (1.13–1.87)	
Jamali Z	2015	25, 1573	17, 20a, 153, 200c, 203, 375, 451, Let-7 g	D	–*			
			193b, 205	D		–*		
			126a	D		–		
			18a, 19a, 20a, 21	U	–*			
			134a, 201	U		–*		
			155	U	–		1.57 (1.22–2.02)	1.00 (0.42–2.6)
Lubov J	2017	116, 8194	17, 20c, 21a, 26a, 195, 203, 218, 375, Lin28B	D	–*			
			34a, 34c-5p, 126a, Let7d, Let-7 g	D		–*		
			205, 451	D	–*	–*		
			9, 18a, 19a, 20a, 23a, 155, 206, 210, 1246	U	–*			
			21	U	–*	–*	1.81 (0.66–2.95)	
			130b-3p, 134, 196a, 213p, 372, 373, 965p, 1413p	U		–*		
Troiano G	2018	15, 1045	16	D		–	1.95 (1.28–2.98)	

(continued)

Table 2.1 (continued)

Author	Year	Number of qualitative studies and number of patients	miR	Level of regulation	Qualitative studies		Quantitative studies	
					OS	DFS	OS HR (95% CI)	DFS HR (95% CI)
			17	D	–*		2.65 (2.07–3.3)	
			20a, 32, 204	D	–			
			101, 125	D	–*			
			21, 155-5p, 196a, 372, 373, 455-5p	U	–			
			29b, 181a, 181b, 1246	U	–*			
Kumarasamy C	2019	50, 6867	34a	D				0.19 (0.01–130.51)
			34c-5p	D			4.36 (2.38–8.00)	
			200b	D			1.19 (0.66–2.18)	
			let-7 g, 17, 20a, 26a, 29c, 34c-5p, 142, 146a, 155, 195, 200b, 203, 212, 218, 300, 375, 451, 548b	D			2.02 (1.42–2.86)	
			18a	U			1.87 (1.05–3.33)	
			21	U			1.59 (1.15–2.19)	
			125b	U			2.3 (0.40–13.40)	

Author	Year	ID	miRNA	D/U	Sign		
			let7a, 9, 18a, 19a, 20a, 21, 93, 100, 125b, 155, 206, 375, 377-3p, 483-5p, 1246	U		1.76 (1.43–2.17)	
			34, 126a, 205	D			2.10 (0.72–6.17)
			21, 21-3p, 96-5p, 130b-3p, 134, 141-3p, 210, 372, 373	U			2.64 (1.92–3.66)
			34, 126a, 205; 21, 21-3p, 96-5p, 130b-3p, 134, 141-3p, 210, 372, 373	D and U			2.60 (1.91–3.51)
Sabarimurugan S	2019	21, 5069	92b	D	+		
			18b, 184, 324-3p, 3188	D	+*		
			29c, 103, 204, 451, 483-5p,744	D	–*		
			18b, 29c, 92b, 103, 184, 204, 324-3p, 451, 483-5p, 744, 3188,	D		0.95 (0.65–1.39)	
			663, let-7c	U	–		
			19b-3p, 18a, 29a, 92a	U	+*		
			10b,17-5p, 21, 22, 572, 638, 1234	U	–*		

(continued)

Table 2.1 (continued)

Author	Year	Number of qualitative studies and number of patients	miR	Level of regulation	Qualitative studies		Quantitative studies	
					OS	DFS	OS HR (95% CI)	DFS HR (95% CI)
			18b, 29c, 92b, 103, 184, 204, 324-3p, 451, 483-5p, 744, 3188	D			0.95 (0.65–1.39)	
			let-7c, 10b, 17-5p, 18a, 19b-3p, 21, 22, 29a, 92a, 572, 638, 663, 1234,	U			1.64 (1.05–2.58)	
			18b, 29c, 92b, 103, 184, 204, 324-3p, 451, 483-5p, 744, 3188; let-7c, 10b, 17-5p, 18a, 19b-3p, 21, 22, 29a, 92a, 572, 638, 663, 1234	D and U			1.19 (0.89–1.60)	

U miR is upregulated, *D* miR is downregulated; − or +reduced or increased, respectively, OS or DFS according to the column; * = statistically significant; **in bold** different miRs whose HRs (95% CI) were pooled together in a meta-analysis, performed to get an overall estimate of their effect when combined. Non-bold miRs are those that, in separate analysis by type of miR, showed the same level of regulation, association and significance in the OS or DFS showed that its upregulation is not significantly associated with a lower probability DFS. The main findings of the systematic reviews and meta-analysis are summarised in Table 2.1.

Discussion and Conclusion

The aim of this review was to summarise evidence about the potential prognostic role of miRNAs as biomarkers in HNC. Some miRNAs were demonstrated to have tumour-suppressing and oncogenic roles according to their level of regulation (upregulation or downregulation) in HNC patients (Lubov et al. 2017).

In this umbrella review, a focus was placed on their role in prognosis; therefore OS and DFS of patients with different miR (assessed together and/or individually) levels of expression were evaluated.

Among all reviews, we found that the most frequently studied miRNR was miR21 which was reported either in the OS and DFS statistical analyses. The OS analysis showed a significant lower prognosis when miR21 (individually or in combination with other miRNAs) was upregulated.

In particular, miR21 has been associated to different cancer types where it is usually upregulated (in both solid and haematological cancers). miR21 was found to be overexpressed in different kinds of tumours, such as glioblastoma, breast, lung, colon, pancreas, prostate, stomach cancer, hepatocellular carcinoma, ovarian cancer, cervical carcinoma, thyroid carcinoma and leukaemia, apart from HNC. The molecular pathways where it is involved are related to oncogenic/oncosuppressive cell signals: the overexpression of miR21 was shown to be associated with cell proliferation, migration, invasion and survival; on the contrary, studies on miR21 knockdown mice have demonstrated that it is linked with apoptosis, repression of cell proliferation and invasion. In humans, its downregulation can be linked to tumour cell growth, survival, chemoresistance, invasion and metastasis (Bourguignon et al. 2009). In particular, it supposed that miR-21, as part of the Stat-3 signal pathway inhibits, by downregulating, the tumour suppressor protein PDCD4 (Programmed Cell Death 4) and promotes the inhibitors of the apoptosis family of proteins (IAP), thus inducing a process of anti-apoptosis, survival and chemoresistance of the cancer cells (Zhou et al. 2014).

In most reviews though, even if the pooled HR was significant in most statistical analyses, its utility as prognostic factor based on the relative strength can be considered moderate or weak as it does not reach a HR > 2 (Hayes et al. 2001). In order to increase the prediction of the prognosis, researchers have recently started to focus on analysis of groups of different miRNAs together. In fact, regarding the OS, two reviews (Kumarasamy et al. 2019; Sabarimurugan et al. 2018) reported a meta-analysis where HRs from different miRNAs expressions were pooled together. One showing that downregulated miRNAs, when taken together, have a pooled HR 2.02 (95% CI 1.43–2.17) (Sabarimurugan et al. 2018), the second one (Kumarasamy et al. 2019) that upregulated miRNAs, when taken together, have a pooled HR 4.64

(95% CI, 1.05–2.58). Furthermore, this latter study (Kumarasamy et al. 2019) highlights that a higher predictive power is obtained in the meta-analysis of DFS of different miRNAs that leads to a pooled HR of 2.64 (95% CI 1.93–3.62) and of 2.14 (95% CI 0.73–6.18), respectively, for upregulated and downregulated miRNAs.

Overall, many limitations occur in our review. First of all, conclusions relative to the level expression (upregulation/downregulation) can be misleading as thresholds regarding the cut-off points are not unequivocally set and they result from primary studies that more widely correspond to mean or median values of the laboratory that performs the analysis. In addition, levels of expression of miRNAs depend on the tissues that are analysed (plasma, serum, tumour tissue) -as miRNAs show a widely variable levels of expression according to the cell type (Witwer and Halushka 2016). Other problems in relation to miRNAs can occur and they range from the challenging techniques of measurement of miRNAs to the duration of the follow-up of patients, characteristics of the sample population (age, ethnicity, tumour stage, tumour type (Fu et al. 2011; Witwer and Halushka 2016). Lastly, as highlighted from the studies included in this review, there is the scarce specificity related to the pathology or tumour site of miRNAs overall.

In conclusion, to our knowledge this study represents the first attempt to summarise published reviews regarding the prognostic role of miRNAs on HNC. Recently, the cumulative effect of sets of miRNAs together has been increasingly studied and they might be stronger predictor of survival than single miRNAs. Eventually, different issues arise from the analysis of miRNAs and the above-mentioned problems still need to be addressed by performing large scale studies to verify and enhance the power of evidence and clinical utility of these biomarkers both individually and in combination.

2.3 Application of Single-Nucleotide Polymorphism-Related Risk Estimates in Identification of Increased Genetic Susceptibility to Non-communicable Diseases

Szilvia Fiatal, Péter Pikó, and Róza Ádány

As the World Health Organization reports (WHO, 2019) of 56.9 million global deaths in 2016, 40.5 million, or 71%, were due to non-communicable diseases (NCDs). Among the leading causes of NCD deaths, the representation of CVDs (17.9 million deaths, 44% of all) was the highest, while diabetes caused another 1.6 million deaths (4% of all).

It is generally accepted that both genetic and environmental/lifestyle risk factors and the complex interplay among them are associated with the aetiology of the so-called cardio-metabolic diseases (e.g. atherosclerosis, T2D, venous thromboembolism (VT) and essential hypertension (EH)). The rapid development of next-generation DNA sequencing technologies and the increasing number of GWAS, as reported in 1.1, have resulted in the identification of more and more disease-related SNPs. Functional polymorphisms (non-synonymous SNPs that change the amino acid sequence) can be used not only to identify new genetic risk markers but also allow health service providers to move further steps towards personalised, as well as precision medicine and prevention. SNPs-based platforms may serve as tools to identify disease susceptibility even before the development of the disease, so on the basis of SNPs-based risk estimates phenotype prevention can be introduced (Torkamani et al. 2018).

Multiple genetic markers (as the SNPs) with usually small effect sizes may be used in combination to generate high(er) effect size. The simultaneous use of the strong risk markers (with or without other non-genetic traditional risk factors) may have the desired discriminatory accuracy to distinguish between high-risk and low-risk subjects and may guide prophylactic and therapeutic decisions. The polygenic/genetic risk score modelling is a method that utilises multiple genetic loci for better prediction of the genetic risk related to a specific trait. Summing up the number of risk alleles with the SNPs examined can be used as an unweighted genetic risk score (GRS) for each individual in the study population. Furthermore, weighted GRS (wGRS) can be constructed considering also the effect size measures (weights) derived from GWA studies.

GRS modelling has become very popular in studies of common diseases. However, medical genetic research, including the use of GRSs, mostly examined populations with European ancestry and just limited numbers were carried out on non-European population (Duncan et al. 2019). Examination of the genetics vulnerability of minorities is essential for the development of effective population-based public health interventions.

The Roma population, which constitutes the largest ethnic minority in Europe, is the main subject of ethnicity-based studies because available data strongly suggest that Roma population groups suffer from poorer health and lower life expectancy. Studies have shown a high prevalence of cardiovascular risk factors (e.g. reduced high-density lipoprotein cholesterol (HDL-C) level, T2D) among Roma independent of the countries where they live; this suggests that, in addition to the environmental/modifiable risk factors, genetic susceptibility may also contribute to the high cardiovascular disease morbidity and mortality burden in the Roma population.

Recently, our research group investigated whether genetic susceptibility contributes to a higher prevalence of the reduced HDL-C level, increased prevalence of T2D and thrombophilia, and lower prevalence of hypertension in the Hungarian Roma (HR) population compared to the Hungarian general (HG) population by utilising the gene scoring approach.

Risk for the Development of Thrombosis

The genetic susceptibility of HR population to venous thrombosis using the similar GRS model suggested by de Haan et al. (de Haan et al. 2012) was investigated and compared to that of the HG population. This study was the first report which showed that a validated SNP panel can be utilised for VT risk estimation in a population-based study, which can foster the translation of genomic results into (public) health practice which is very much required because, so far, only limited evidence is available (Fiatal and Ádány 2017). It was demonstrated that the HR population was significantly more susceptible to VT using genetic risk scoring models (Fig. 2.1a). The mean of the GRS was 41.83 ± 5.78 in the HR and was 41.04 ± 6.04 in the HG population ($p = 0.001$). Interventions aiming to improve the Roma's cardiovascular health and to identify subgroups of high-risk individuals need to consider that this ethnic population has an increased genetic propensity to VT (Fiatal et al. 2019).

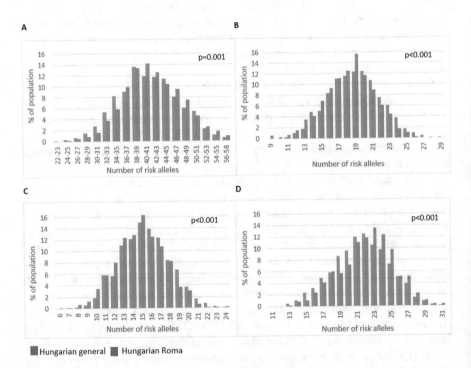

■ Hungarian general ■ Hungarian Roma

Fig. 2.1 Distribution of unweighted genetic risk scores for the development of thrombosis (**a**, based on 48 SNPs), hypertension (**b**, based on 20 SNPs), T2DM (**c**, based on 16 SNPs) and atherosclerosis (**d**, based on 21 SNPs) in Hungarian general and Roma populations. Unweighted (poly) genic risk score (GRS) is calculated on the basis of the number of risk alleles carried. Thus, "0" indicates absence of the risk allele, while risk allele homozygotes are coded as genotype "2" and heterozygotes as genotype "1". In GRS, the model makes the assumption that all markers are of equal importance in calculating genetic susceptibility

Risk for the Development of Hypertension

A study on comparing the genetic susceptibility to EH—one of the well-established risk factor for an adverse cardiovascular outcome—in the HR population to that of the HG population was designed. The intention was to investigate whether disparities exist in the cumulative risk allele between the two populations, which may—at least to a certain extent—elucidate ethnic differences in EH prevalence. It was shown that the average GRS was significantly lower among the HR population compared to the HG (Fig. 2.1b), even after adjusting for the effects of possible confounders (HR: $18.25 \pm$ SD 2.97 and HG: $18.98 \pm$ SD 3.05, $p < 0.001$). The Roma seems to be genetically less susceptible to EH than the general one. These results support preventive efforts to lower the risk of developing EH by focusing on the harmful environmental or behavioural factors rather than their genetic propensity (Soltész et al. 2020).

Risk for the Development of Type 2 Diabetes Mellitus

It is generally accepted that, in addition to lifestyle/environmental conditions, genetic factors also have a considerable effect on the development of T2D. The age- and gender-adjusted heritability for elevated fasting blood glucose level was estimated to be 38%. Estimates for the heritability of T2D come from a variety of population, family, and twin-based studies and range from 20% to 80%.

Numerous studies have investigated the applicability of GRS calculation for susceptibility to T2D. A significant proportion of these studies reported low predictive value for models that rely solely on genetic factors (area under curve (AUC) was from 0.550 to 0.680), while the impact of environmental and lifestyle factors (AUC of conventional model: 0.614–0.959) was found to be significant. However, the weight of genetic factors is not negligible, as the predictive value of combined models (AUC of combined models: 0.631–0.963) has always outstripped that of conventional models. This phenomenon was also observed in the HR population, where the increased prevalence of prediabetes and known T2D—27.09% in the HR vs. 15.56% in HG; $p < 0.001$ (Kosa et al. 2015)—was clearly associated with ethnicity-related environmental and lifestyle characteristics and not with genetic factors (Fig. 2.1c) (Werissa et al. 2019).

Risk for the Development of Atherosclerosis (Reduced HDL-C Level)

The atherogenic plasma lipoprotein profile is one of the most important risk factor/predictor for the development of CVDs. The HDL-C is the key component involved in reverse cholesterol transport and the transfer of cholesteryl esters between lipoproteins. The level of HDL-C is inversely associated with the risk of coronary heart diseases and is interpreted as a key indicator of predicting CVDs risk. In a large-scale epidemiological study (Tehran Lipid and Glucose Study), the age- and sex-adjusted heritability for HDL-C as a continuous trait was 46% and for low HDL-C (<1.03 mmol/l in males and < 1.29 mmol/l in females) was estimated to be 40%.

In our previous study, it was found that the prevalence of reduced plasma HDL-C level was significantly more frequent in all age groups of the HR population compared to the general one (HR: 54.2% vs HG: 36.06%, $p = 0.029$) (Kosa et al. 2015). Both the GRS and wGRS (based on 18 HDL-C related SNPs) were significantly higher in the HR population compared to the HG population (GRS: 22.2 ± 3.2 vs. 21.5 ± 3.3; wGRS: 0.57 ± 0.1 vs. 0.53 ± 0.1; $p < 0.001$) (Fig.2.1d), and showed significant association with the HDL-C level as a continuous trait and reduced HDL-C status in both populations (Pikó et al. 2017).

2.4 Polygenic Risk Score for Data-Driven Health Care in the Big Data Age

Giordano Bottà

The UK Biobank

Beginning this subchapter without presenting the UK Biobank wouldn't give justice to the half million participants and to the hundreds of people who worked at the largest prospective genomics study to date. The UK biobank made available to research institutions and private companies an astonishing amount of data, comprising the genotype at ~800 K SNPs of about 500.000 people, coupled with their electronic health records, providing deep and curated phenotypes. The UK Biobank is reaching full 10 years of follow-up in 2020 (Bycroft et al. 2018).

In 2019, it was announced that all the participants will be whole genome sequenced thanks to a collaborative effort among the UK government investing £50 million, the Wellcome Trust charity investing £50 million and four pharmaceutical companies: Amgen, AstraZeneca, GlaxoSmithKline and Johnson & Johnson investing a total of £100 million (https://www.ukbiobank.ac.uk). The magnitude and scale of the data being gathered is unprecedented, having the potential to fundamentally improve and change our understanding of the interaction between human genome and diseases.

A radical change of paradigm in the genomic industry is also observed, caused by the public availability of this large amount of data. The main asset of private companies in the healthcare space is not anymore the proprietary data that a company was able to collect, but the expertise in developing algorithms and software to mine large datasets, now available to everyone. This change of paradigm was triggered by the impossibility of a single private company to reach the scale of data and the level of phenotype curation that can be achieved by a heavily digitalised public health system investing in genomic research. This change of paradigm is improving population and patient health. A large number of startups with limited resources, but with high technical skills and medical knowledge, are now able to bring to the market new products to improve people's life, competing with large and well-established companies.

Polygenic Risk Score

One of the main opportunities enabled by the UK Biobank dataset is to validate the power of genetic variants in predicting the development of complex diseases. By definition, complex diseases are caused by a combination of multiple factors, genetic and environmental (Hunter, 2005). In the last decade, the understanding of the genetic architecture of complex diseases has been the focus of a large number of studies using a method called GWAS as already reported in 1.1 and 1.3 (Visscher et al. 2017). GWAS are hypothesis-free studies, since the entire genome is analysed without prioritising specific genetic variants, even if they have been found associated with the disease in previous studies. GWASs performed on complex diseases usually found multiple genetic variants with significant p-values highlighting the polygenicity of these kind of diseases that are caused by a combination of common variants and not only by a single gene mutation. The disease-associated effect size of hundreds or thousands of genetic variants identified through a GWAS can be summed in a single number called polygenic risk score (PRS), as briefly outlined in 1.3 (Torkamani et al. 2018). PRS is computed for an entire population, then the disease risk in the tail of the distribution are compared with the reminder to identify sub-populations at higher risk that potentially need more aggressive and urgent preventive strategies. A long-lasting open question was if common genetic variants used to build PRSs were able to predict risk of developing complex diseases with enough power to be used in a clinical setting. The lack of large enough dataset, able to allow an unbiased PRS validation, hampered this question to be answered for many years.

The UK biobank allowed for the first time a robust validation of predictive models for complex disease based on PRS thanks to its hundreds of thousands of samples and the possibility to be divided into two large validation and testing datasets. At the same time the development of new powerful algorithm to model GWAS summary statistics (Privé et al. 2019) completed this exciting picture where predictive models to identify risk for common diseases have now clinical utility, allowing personalisation of preventive strategies.

Human diversity is beautiful in its essence and at the same time very complicated from a scientific point of view. PRS discovered in European populations cannot be directly transferred to other ethnicities. Association patterns between genetic variations of different ancestry can vary substantially, avoiding the portability of GWAS summary statistics to build PRS between ancestry groups. This is the case for African populations, where we observe the highest genetic diversity and consequently a different genetic variants association pattern compared to European populations. Unfortunately to date there is an under-representation of non-European populations in GWAS, limiting the development of ancestry-specific PRS. This is a priority issue that must be tackled to avoid health disparities between ancestry groups (Martin et al. 2019).

In the last decade the enthusiasm in using genetics to personalise prevention for complex diseases was tempered by the impossibility to identify populations at enough higher risk to guide personalised clinical intervention. In 2018, a study published in Nature Genetics built new hope in using genetics in preventive clinical model since it showed that PRS can identify individuals with risk equivalent to those people already considered at high risk in a clinical setting, such as the familiar hypercholesterolaemia carriers (Khera et al. 2018). Moreover, recent data showed that PRS is also able to stratify risk in single gene mutation carrier such as *BRCA 1/2*. This PRS application is crucial to allow women to make informed decision in case of the presence of a *BRCA* mutation: the overall risk can vary substantially due to the polygenic background, from 13% to almost 76% (Fahed et al. 2019). Knowing to be a single gene mutation carrier is evidently not enough, PRS plays a crucial role in defining mutation penetrance and it can also be analysed with a non-expensive genetic test. The coming decade will likely see a rapid implementation in the clinical setting of predictive models based on the newly developed PRS able to identify a large fraction of European population at clinical significant risk of developing complex disease.

New Challenges in Preventive Medicine

Papers published in the last 2 years highlighted a new scenario in preventive medicine where the use of common variants (i.e. SNPs) allows the identification of a large number of individuals at increased risk who needs personalised preventive strategies. Genomic tests for risk assessment have now a public health value not only to identify risk in specific populations already considered at increased risk, such as people with a first relative family history, but also for the general population as a screening tool. For example, women with high PRS for breast cancer can reach the same risk of a 50-year-old woman many years before, while at the same time women with low PRS reach the risk of a 50-year-old woman at more than 70 years of age (Mavaddat et al. 2019). Several public health systems recommend preventive mammographies for breast cancer starting at age of 50. Such recommendations do not take into account the underlying individual specific genetic risk, whose effectiveness in terms of health outcomes has the potential to improve by incorporating this additional and relevant genetic information.

A data-driven preventive strategy opens new challenges and opportunities. There are two main challenges: education and cost-effectiveness, which are also further addressed in Chaps. 2 and 3.

Concerning education, genetics was always considered as an exceptional science, where variations in our genome were known to be able to determine our destiny in an unchangeable fashion. This is in part justified by the early discoveries of single gene mutations with large penetrance. On the other hand, the understanding of the genetic underpinning of complex diseases shifted the deterministic view towards a model where genetics is considered as a risk factor, composed of many common variants with small effect. Genetic risk factors act in conjunction with traditional risk factors and the existence of this complex interplay should be taught to both the general population and healthcare professionals. There is also a cultural problem that hampers the adoption of genomic test for common diseases in primary prevention. Often physicians consider a genetic biomarker in a binary way; if the risk conferred by the genetic variation approaches 100%, it is considered to be useful, otherwise as useless. But if we think carefully about that, we realise that in medicine physicians use several risk factors that are not deterministic and still have a very useful role in risk modelling. For example, blood cholesterol has lower power than PRS in predicting the development of coronary artery disease, yet it is considered as a major player in absolute risk modelling.

Fundamental is to educate since the very first genetic courses that single gene mutation with very high penetrance are not the norm, rather an exception, and genetics can be considered as a risk factor in the majority of the cases, with higher or lower predictive power depending on the diseases. What must be stressed is the need to provide evidence that incorporating a genetic risk factor into a risk model will result in an improvement of the existent model and in a favourable clinical measurable outcome.

Concerning cost-effectiveness, this second challenge requires the development of complex economics models that vary at a state and regional level. In Europe, the cost for curing a person with a disease varies significantly, and the benefit of using genomic screening for complex diseases must be evaluated at a very fine scale. Unfortunately, the fragmentation present between European countries and also within, at a regional level, renders very difficult to perform broadly representative analysis. This problem should be addressed at a local level performing small pilot studies on populations representative of the local genetic and cultural diversity. Preliminary data are extremely encouraging showing improvement of risk modelling with the integration of PRS, and considering the low price of a genotype array needed to perform the analysis won't be long before we'll see a mass adoption of PRS-based risk modelling as a national level screening tool.

Market Needs

Clinical relevant DNA-based risk factors such as PRS will be undoubtedly routinely used in clinical setting. Computing PRS is not trivial for an average genetic laboratory as it requires a series of complex bioinformatics steps, from data harmonisation, data enrichment, data curation, quality control and finally PRS calculation.

Genetics laboratories are usually equipped with desktop personal computer that have not enough computation power to perform these analyses. Cloud computing is the easiest solution to allow mass adoption of PRS analysis. The term cloud indicates a net of computers remotely located that can be switched on and off from a normal personal computer, the only requirement is an internet connection.

There is also a substantial saving in using the cloud. Software engineers can write codes that allow the elastic use of cloud computing, opening bigger and most expensive machines only when needed or using the smaller ones to optimise costs when lighter analysis is performed. In the past, the use of cloud computing in the healthcare sector was viewed with suspicion because it requires data to be moved outside from a hospital or a healthcare facility. The newly enforced European data privacy protection law, called General Data Protection Regulation (GDPR), forced stakeholders in building more trusted cloud computing, because each data movement must be documented and explicitly accepted by the data owner. For this reason, there are very few cases were cloud computing is not accepted and an on-premise solution is required. The downside of an on-premise solution is that healthcare facilities must adopt big computational power that are costly and difficult to maintain for a non-IT expert.

Certifications are another important aspect to take into consideration. Softwares and analysis pipeline embedded in them can be certified as a medical device, requiring a very in-depth documentation with clinical use cases, that must prove the analytical validity and clinical utility.These challenges are opening a very big market to products built by a collaborative effort between software engineers and geneticists to create digital products able to fulfill technical and clinical needs. Future technological innovations in genome sequencing and cloud computing will also accelerate the adoption of new PRS-based risk models.

In conclusion, the combination of single gene mutation and polygenic background in defining an individual genetic risk factor brings the potential to shape a bright future, where robust and unbiased risk assessment might lead to precise preventive medicine effective both at a personal and population level.

References

Babu, J. M., Prathibha, R., Jijith, V. S., et al. (2011). A miR-centric view of head and neck cancers. *Biochimica et Biophysica Acta, Reviews on Cancer, 1816*, 67–72.

Bourguignon, L. Y. W., Spevak, C. C., Wong, G., et al. (2009). Hyaluronan-CD44 interaction with protein kinase Cε promotes oncogenic signaling by the stem cell marker nanog and the production of microRNA-21, leading to down-regulation of the tumor suppressor protein PDCD4, anti-apoptosis, and chemotherapy resistance in breast tumor cells. *Journal of Biological Chemistry, 284*(39), 26533–26546.

Bycroft, C., Freeman, C., Petkova, D., et al. (2018). The UK Biobank resource with deep phenotyping and genomic data. *Nature, 562*(7726), 203–209.

Carleton, M., Cleary, M. A., & Linsley, P. S. (2007). MicroRNAs and cell cycle regulation. *Cell Cycle (Georgetown, Tex.), 6*(17), 2127–2132.

Chen, R., Xia, L., Tu, K., et al. (2018). Longitudinal personal DNA methylome dynamics in a human with a chronic condition. *Nature Medicine, 24*, 1930–1939.

Chin, D., Boyle, G. M., Porceddu, S., Theile, D. R., et al. (2006). Head and neck cancer: Past, present and future. *Expert Review of Anticancer Therapy, 6*, 1111–1118.

De Haan, H. G., Bezemer, I. D., Doggen, C. J., et al. (2012). Multiple SNP testing improves risk prediction of first venous thrombosis. *Blood, 120*, 656–663.

Duncan, L., Shen, H., Gelaye, B., et al. (2019). Analysis of polygenic risk score usage and performance in diverse human populations. *Nature Communications, 10*, 3328.

Fahed, A. C., Wang, M., Homburger, J. R., et al. (2019). Polygenic background modifies penetrance of monogenic variants conferring risk for coronary artery disease, breast cancer, or colorectal cancer. *medRxiv*, 19013086.

Fiatal, S., & Adany, R. (2017). Application of single-nucleotide polymorphism-related risk estimates in identification of increased genetic susceptibility to cardiovascular diseases: A literature review. *Frontiers in Public Health, 5*, 358.

Fiatal, S., Piko, P., Kosa, Z., et al. (2019). Genetic profiling revealed an increased risk of venous thrombosis in the Hungarian Roma population. *Thrombosis Research, 179*, 37–44.

Fu, X., Han, Y., Wu, Y., et al. (2011). Prognostic role of microRNA-21 in various carcinomas: A systematic review andmeta-analysis. *European JournalL of Clinical Investigation, 41*(11), 1245–1253.

Hayes, D. F., Isaacs, C., & Stearns, V. (2001). Prognostic factors in breast cancer: Current and new predictors of metastasis. *Journal of Mammary Gland Biology and Neoplasia, 6*(4), 375–392.

Hunter, D. (2005). Gene-environment interactions in human diseases. *Nature Reviews Genetics, 6*(4), 287–298.

Jamali, Z., Asl Aminabadi, N., Attaran, R., et al. (2015). MicroRNAs as prognostic molecular signatures in human head and neck squamous cell carcinoma: A systematic review and meta-analysis. *Oral Oncology, 51*(4), 321–331.

Kengne, A. P., Beulens, J. W., Peelen, L. M., et al. (2014). Non-invasive risk scores for prediction of type 2 diabetes (EPIC-InterAct): A validation of existing models. *The Lancet Diabetes and Endocrinology, 2*, 19–29.

Khera, A. V., Chaffin, M., Aragam, K. G., et al. (2018). Genome-wide polygenic scores for common diseases identify individuals with risk equivalent to monogenic mutations. *Nature Genetics, 50*(9), 1219–1224.

Kosa, Z., Moravcsik-Kornyicki, A., Dioszegi, J., et al. (2015). Prevalence of metabolic syndrome among Roma: A comparative health examination survey in Hungary. *European Journal of Public Health, 25*, 299–304.

Kumarasamy, C., Madhav, M. R., Sabarimurugan, S., et al. (2019). Prognostic value of miRNAs in head and neck cancers: A comprehensive systematic and meta-analysis. *Cell, 8*(8), 772.

Liu, J., Carnero-Montoro, E., van Dongen, J., et al. (2019). An integrative cross-omics analysis of DNA methylation sites of glucose and insulin homeostasis. *Nature Communications, 10*, 2581.

Liu, J., Semiz, S., van der Lee, S. J., et al. (2017). Metabolomics based markers predict type 2 diabetes in a 14-year follow-up study. *Metabolomics, 13*, 104.

Lubov, J., Maschietto, M., Ibrahim, I., et al. (2017). Meta-analysis of microRNAs expression in head and neck cancer: uncovering association with outcome and mechanisms. *Oncotarget, 8*(33), 55511–55524.

Mahajan, A., Taliun, D., Thurner, M., et al. (2018). Fine-mapping type 2 diabetes loci to single-variant resolution using high-density imputation and islet-specific epigenome maps. *Nature Genetics, 50*, 1505–1513.

Martin, A. R., Kanai, M., Kamatani, Y., et al. (2019). Clinical use of current polygenic risk scores may exacerbate health disparities. *Nature Genetics, 51*(4), 584–591.

Mavaddat, N., Michailidou, K., Dennis, J., et al. (2019). Polygenic risk scores for prediction of breast cancer and breast cancer subtypes. *Am J Hum Genet, 104*(1), 21–34.

Piko, P., Fiatal, S., Kosa, Z., et al. (2017). Genetic factors exist behind the high prevalence of reduced high-density lipoprotein cholesterol levels in the Roma population. *Atherosclerosis, 263*, 119–126.

Privé, F., Vilhjálmsson, B. J., Aschard, H., et al. (2019). Making the most of clumping and thresholding for polygenic scores. *Am J Hum Genet, 105*(6), 1213–1221.

Sabarimurugan, S., Madurantakam Royam, M., Das, A., et al. (2018). Systematic review and meta-analysis of the prognostic significance of miRNAs in melanoma patients. *Molecular Diagnosis & Therapy, 22*(6), 653–669.

Shea, B. J., Reeves, B. C., Wells, G., et al. (2017). AMSTAR 2: A critical appraisal tool for systematic reviews that include randomised or non-randomised studies of healthcare interventions, or both. *BMJ (Online), 358*, j4008.

Soltesz, B., Piko, P., Sandor, J., et al. (2020). The genetic risk for hypertension is lower among the Hungarian Roma population compared to the general population. *PLoS One, 15*(6), e0234547.

Tabák, A. G., Jokela, M., Akbaraly, T. N., et al. (2009). Trajectories of glycaemia, insulin sensitivity, and insulin secretion before diagnosis of type 2 diabetes: an analysis from the Whitehall II study. *Lancet, 373*, 2215–2221.

Thomas, G. R., Nadiminti, H., & Regalado, J. (2005). Molecular predictors of clinical outcome in patients with head and neck squamous cell carcinoma. *International Journal of Experimental Pathology, 86*(6), 347–363.

Torkamani, A., Wineinger, N. E., & Topol, E. J. (2018). The personal and clinical utility of polygenic risk scores. *Nature Reviews. Genetics, 19*, 581–590.

Troiano, G., Mastrangelo, F., Caponio, V. C. A., et al. (2018). Predictive Prognostic Value of Tissue-Based MicroRNA Expression in Oral Squamous Cell Carcinoma: A Systematic Review and Meta-analysis. *Journal of Dental Research, 97*(7), 759–766.

Visscher, P. M., Wray, N. R., Zhang, Q., et al. (2017). 10 years of GWAS discovery: Biology, function, and translation. *Am J Hum Genet, 101*(1), 5–22.

Wahl, S., Drong, A., Lehne, B., et al. (2017). Epigenome-wide association study of body mass index, and the adverse outcomes of adiposity. *Nature, 541*, 81–86.

Wang, T. J., Larson, M. G., Vasan, R. S., et al. (2011). Metabolite profiles and the risk of developing diabetes. *Nature Medicine, 17*, 448–453.

Wellcome Trust Case Control Consorcium. (2007). Genome-wide association study of 14,000 cases of seven common diseases and 3,000 shared controls. *Nature, 447*, 661–678.

Werissa, N. A., Piko, P., Fiatal, S., et al. (2019). SNP-based genetic risk score modeling suggests no increased genetic susceptibility of the roma population to type 2 diabetes mellitus. *Genes (Basel), 10*(11). pii: E942. https://doi.org/10.3390/genes10110942.

WHO. (2019). World Health Organization: NCD mortality and morbidity. Global Health Observatory data. [Online]. Retrieved form https://www.who.int/gho/ncd/mortality_morbidity/en. Accessed 3 Dec 2019.

Witwer, K. W., & Halushka, M. K. (2016). Toward the promise of microRNAs – Enhancing reproducibility and rigor in microRNA research. *RNA Biology, 13*, 1103–1116.

Zhou, X., Ren, Y., Liu, A., et al. (2014). WP1066 sensitizes oral squamous cell carcinoma cells to cisplatin by targeting STAT3/miR-21 axis. *Scientific Reports, 4*.

Chapter 3
Evaluation of Predictive Genomic Applications

Paolo Villari, Erica Pitini, Elvira D'Andrea, and Annalisa Rosso

3.1 The Health Technology Assessment Approach for the Evaluation of Genetic/Genomic Applications

Erica Pitini

Introduction

The rapid proliferation of tests for genetic disorders has made the routine evaluation of their benefits, risks and limitations of paramount importance, so that only those tests with proven benefits are implemented in clinical and public health practice. This is essential to avoid the harmful effects on patient health and management, and avoidable waste of resources, that an ineffective test would represent. The need for a robust evaluation strategy for genetic tests has been recognised and emphasised across Europe and worldwide (Boccia et al. 2014; Burke et al. 2002).

P. Villari (✉) · E. Pitini · A. Rosso
Department of Public Health and Infectious Diseases, Sapienza University of Rome, Rome, Italy
e-mail: paolo.villari@uniroma1.it; erica.pitini@uniroma1.it; annalisa.rosso@uniroma1.it

E. D'Andrea
Department of Public Health and Infectious Diseases, Sapienza University of Rome, Rome, Italy

Division of Pharmacoepidemiology and Pharmacoeconomics, Brigham and Women's Hospital, Harvard Medical School, Boston, MA, USA
e-mail: edandrea@bwh.harvard.edu

S. Boccia et al. (eds.), *Personalised Health Care*, SpringerBriefs in Public Health, https://doi.org/10.1007/978-3-030-52399-2_3

33

However, the evaluation of genetic tests is not a simple matter. Although several evaluation frameworks have been proposed in recent years, on closer inspection, they mainly rely on two well-known evaluation approaches, i.e. the ACCE (analytic validity, clinical validity, clinical utility, and ethical, legal, and social implications) model and the Health Technology Assessment (HTA) process (Pitini et al. 2018).

The ACCE Model

The ACCE model was developed in the early 2000s by the Centers for Disease Control and Prevention (CDC)'s Office of Genomics and Disease Prevention, specifically for the evaluation of genetic tests (CDC 2010). It derives its name from the evaluation dimensions used, i.e. analytic validity, clinical validity, clinical utility, and ethical, legal and social implications. Analytic validity is defined as the ability of the test to accurately and reliably measure the genotype of interest, while clinical validity is the ability of the test to accurately and reliably detect or predict a clinical condition. Clinical utility is defined as the likelihood that the test will lead to improved outcomes. Notably, the ACCE model also includes in clinical utility some consideration of contextual issues, such as economic benefits and organisational aspects (facilities, personnel, etc.). Finally, ethical, legal and social implications are defined as the safeguards and impediments relating to the impact of test results on stigmatisation, discrimination, privacy and confidentiality, and personal, family or social issues (CDCP 2010). From its inception, the ACCE model has been adopted and adapted by various entities both in the United States and worldwide. In 2004, several aspects of the ACCE model were used in the Evaluation of Genomic Applications in Practice and Prevention (EGAPP) initiative, which aimed to assess genetic tests and make recommendations for their use in clinical and PH practice (Teutsch et al. 2009). In some countries (e.g. the UK, Germany, Andalusia and Australia), the ACCE model was further developed to guide the introduction of new genetic tests into their national health systems (Pitini et al. 2018); it also inspired the Clinical Utility Gene Card, i.e. the assessment tool of EuroGentest, a project funded by the European Commission to harmonise the process of genetic testing across Europe (Schmidtke and Cassiman 2010). In an initiative supported by the PHG Foundation, the ACCE model has also been expanded to incorporate the use of health quality measures in the evaluation of a genetic test and associated services (Burke and Zimmern 2007). The above are only a few examples of applications of the ACCE model; its success is due to the fact that standard methods for the assessment of health technologies cannot adequately be applied to genetic tests, which have specific aspects that must be understood, particularly analytic validity and clinical validity, before the tests can be considered suitable for clinical use.

The HTA Model

The HTA model is also widely used for the evaluation of genetic tests. In contrast to the ACCE approach, it was developed to cover all health technologies, but some attempts have been made to adapt it for the evaluation of genetic tests. One of the most remarkable is the framework proposed to guide the public coverage of new predictive genetic tests in Ontario (Canada) (Giacomini et al. 2003). The main innovation of this framework is the unit of analysis: what must be assessed for coverage is not just the laboratory test, but the whole testing service, i.e. the laboratory technology, plus the target population, plus the clinical context. The latter includes clinical objectives (e.g. screening, diagnosis, treatment planning) and route of access (e.g. private market, by referral). The criteria proposed to assess the genetic testing service are intended purpose, effectiveness, additional effects, aggregate costs, demand and cost-effectiveness. This framework is a good example of the service delivery approach that, according to Battista, is a peculiarity of HTA (Battista 2006). Nevertheless, applications of HTA to genetic tests have not yet attained the complexity and comprehensiveness of general HTA frameworks. For example, the European reference framework for HTA, the EUnetHTA HTA core model, requires a much more detailed analysis of organisational and economic aspects than is currently used in HTA evaluations of genetic tests. Moreover, this model considers additional evaluation dimensions such as health problem and current use of technology; description and technical characteristics of technology; safety; clinical effectiveness; ethical aspects; organisational aspects; patients and social aspects; and legal aspects (EUnetHTA Joint Action 2013).

The Need of a More Comprehensive Evaluation Framework?

Which approach should then be preferred to address the full spectrum of issues relating to the evaluation of genetic tests? Since there are ways in which the evaluation of genetic tests is relatively unique and since decision makers are constantly challenged by equity and resource constraints, a combination of the ACCE model, which is well suited to genetic tests, and the HTA process, which allows a more systematic analysis of service delivery, might represent the best strategy based on current findings.

Such an integrated approach has recently been realised by the Department of Public Health and Infectious Diseases of Sapienza University (Table 3.1) (Pitini et al. 2019). This new framework is distinguished by a dual focus on both the genetic test and its delivery models. The first section of the framework, "Genetic test", is devoted to the technical and clinical value of a genetic test and is mostly based on most of the ACCE evaluation dimensions (analytic validity, clinical validity, clinical

Table 3.1 Evaluation framework of genetic tests developed by Sapienza University of Rome (Pitini et al. 2019)

Sections	Evaluation dimensions

I. Genetic test	• Test and clinical condition overview
	  – Clinical condition
	      Clinical presentation and pathophysiology
	      Genetic background
	      Public health impact
	  – Genetic test
	      General features
	      Technical features
	      Clinical context
	• Analytic validity
	  – Analytic sensitivity
	  – Analytic specificity
	  – Accuracy
	  – Precision
	  – Robustness
	  – Laboratory quality control
	• Clinical validity
	  – Scientific validity
	  – Test performance
	      Clinical sensitivity and specificity
	      Positive and negative predictive value
	      Modifiers
	• Clinical utility
	  – Available interventions
	  – Efficacy
	  – Effectiveness
	  – Safety
	• Personal utility

II. Delivery models	• Delivery models overview
	  – Healthcare programmes
	  – Level of care
	  – Patient pathway
	• Organisational aspects
	  – Expected demand
	  – Resources management
	  – Other organisational requirements
	      Education of professionals, patients and citizens
	      Information dissemination to professionals, patients and citizens
	      Cooperation, communication and coordination
	      Quality assurance, monitoring and control
	  – Barriers to implementation
	• Economic evaluation
	• Ethical, legal and social implications
	• Patient perspective

III. Research priorities	• Evidence gaps and research priorities
IV. Decision points	• Net benefit • Cost-effectiveness • Feasibility

(continued)

Table 3.1 (continued)

Source: Reproduced from Pitini E, D'Andrea E, De Vito C, et al. (Pitini et al. 2019) A proposal of a new evaluation framework towards implementation of genetic tests. PLOS ONE 14(8): e0219755, Table 4. The proposed evaluation framework. https://doi.org/10.1371/journal.pone.0219755, licensed under the terms of the Creative Commons Attribution License (https://creativecommons.org/licenses/by/4.0/)

utility), with the addition of a single extra dimension, viz. personal utility. This addition was necessary because the framework defines clinical utility in its narrowest sense, i.e. an improvement in health outcomes due to the test and the subsequent clinical interventions. Although health outcomes are undoubtedly the most critical factor in setting priorities for public health, especially in publicly funded healthcare systems, the personal utility dimension is important because it addresses the non-medical benefits the test may have on patients. These include a better understanding of the disease and an improved ability to make relevant life decisions (e.g. enabling reproductive choices or reducing risk-taking behaviour) (Kohler et al. 2017). The second section, "Delivery models", is devoted to the analysis of service delivery models and is mostly based on the HTA approach, particularly the EUnetHTA HTA core model. A delivery model for the provision of genetic tests is defined as the broad context in which genetic tests are offered to individuals and families with or at risk of genetic disorders (Unim et al. 2019). This builds on and extends the testing service concept introduced by the Ontario framework. A delivery model should comprise three elements: the healthcare programme, i.e. the set of health interventions preceding, following and including the genetic test, for a specific target population and with a specific health purpose; the patient pathways, i.e. the patient flow through different professionals from the point of access to the genetic test to the diagnosis and treatment of the genetic disorder; and the level of care (e.g. primary or specialist care level) in which the provision of the genetic healthcare programme is integrated and coordinated (Pitini et al. 2019). The evaluation dimensions to be considered in this section are organisational aspects, economic evaluation, ethical, legal, and social implications (ELSI) and patient perspective. It must be noted that the introduction of patient perspective as an independent evaluation dimension is relatively new in the evaluation of genetic tests and health technologies in general. It is a response to the increasing international interest in patient-centred care, where an individual's specific health needs and desired health outcomes should be regarded as the driving force behind healthcare decisions, quality measurements and resource allocation (NEJM Catalyst 2017). Thus, the patient's "perceived value" of a technology needs to be acknowledged in the evaluation process.

The value of considering delivery models when collecting economic, organisational, ELSI and patient-perspective evidence can be better explained using the example of *BRCA1/2* genetic test delivery options. As we pointed out previously (Pitini et al. 2019), *BRCA1/2* genetic tests could be delivered in a population screening programme, where healthy women invited for mammography may be advised to take a *BRCA1/2* test on the basis of their family history. In this case, local healthcare units would need to coordinate active call of patients, risk assessments, genetic counselling, preventive or treatment pathways, cascade screening, etc. A *BRCA1/2* genetic test could also be delivered in a specialist care setting, such as an oncology clinic, where women with breast cancer may be advised to take the test on the basis of their personal and family history. In this case, active call to recruit the general population is not needed, and counselling and other interventions will probably take place in a specialist context. These two different delivery models will definitely have different costs, benefits, organisational issues, ethical concerns and impact on patient perspective. Thus, they will prove more or less appealing for decision makers.

However, this approach is not free from limitations, the main one being the context dependence of findings. In fact, the analysis of the delivery models, and particularly the assessment of organisational and economic aspects, depends on contextual factors such as the nature of the healthcare system and the level at which decisions are taken, e.g. regional, departmental or hospital. Nevertheless, decision makers need to consider these contextual factors alongside technical and clinical outcomes to secure an efficient and equitable allocation of healthcare resources and services.

Another issue affecting the evaluation of genetic tests is the lack of evidence, especially that relating to the clinical value and implementation of a genetic test. In fact, while gene discovery studies are carried out relatively rapidly, comparatively translation studies (e.g. clinical trials, observational studies, implementation research, outcome research) often lag behind (Khoury 2017). Thus, finding answers to all the questions posed by the evaluation dimensions in a comprehensive evaluation framework is likely to be difficult, at least with the information currently available. Some genetic test evaluation frameworks, including the Sapienza framework, attempt to deal with the evidence gaps uncovered during the assessment process by formulating priorities to guide relevant future research.

Finally, we must consider that, although evidence collection arguably represents the main focus of any technology assessment, it is only one step of a multi-step process, which should also include priority setting and appraisal. As the resources available for HTA are inadequate to address the increasing number of available genetic tests, a valid method of setting assessment priorities is needed. Several methods for priority setting are available for standard health technologies, but their use for genetic tests would require validation. Furthermore, as the main intention of HTA is to inform decisions, the large body of evidence collected through an evaluation framework needs to be appraised so that final recommendations on the adoption and use of a genetic test can be made. Even here, several processes and criteria for appraisal have been developed for standard technologies, and some existing genetic test evaluation frameworks have suggested criteria for making recommendations based on the evidence collected. For example, the Sapienza framework proposed the synthesis of the collected evidence into three practical points for decision, namely the net benefit, cost-effectiveness and feasibility of a particular test. Nevertheless, such attempts need to be improved if they are to facilitate the difficult task of establishing the weight of many different aspects in the overall assessment of a genetic test.

3.2   The Economic Evaluation of Genomic Applications

Elvira D'Andrea

Economic Evaluation in Healthcare

Economic evaluation is a tool used to efficiently allocate resources to medical interventions with the ultimate goal of maximising health outcomes for a population. This tool can be used not only to compare and generate decisions on alternative treatments, medical devices or health services (i.e. clinical research), but also on alternative delivery programmes for implementing treatments, devices or health services (i.e. dissemination research). Even more broadly, economic evaluation can also be applied to alternative strategies for implementing delivery programmes (i.e. implementation research) (D'Andrea et al. 2015).

At first glance, economic evaluation might appear to be mainly focused on cost-containment. However, economic evaluation may or may not save money, since it is only one of many elements involved in healthcare decisions. The key aim of economic evaluation is, rather, to improve the overall *value* of healthcare, which is defined as health outcomes achieved per unit of cost, while considering the individuality of patients and their preferences for care (Porter 2010). In fact, economic evaluation can unearth underused interventions, substitute delivery programmes or new implementation strategies that might increase costs but represent good value for money considering the health outcomes achieved. It can also reveal overused interventions, not particularly effective delivery programmes or outdated implementation strategies that result in poor overall value for money (Neumann et al. 2005).

The two main characteristics of an economic evaluation are (i) a comparative analysis of two or more alternatives (e.g. treatments, devices, services, delivery programmes, implementation strategies), and (ii) an analysis of the outcomes of the interventions, taking into account both the costs and consequences of each alternative. In healthcare, the consequences of an intervention are measured in terms of health effects (e.g. years of life gained, cases of disease or adverse events avoided) and must be comparable across alternatives. The most common method for evaluating health effects and the cost of alternatives is cost-effectiveness analysis. A special case of cost-effectiveness analysis is cost-utility analysis, in which the health effects are measured using quality-adjusted life-years (QALYs). According to recommendations made by the Second Panel on Cost-Effectiveness in Health and Medicine (2016), supported by the US Public Health Service, and similar recommendations made by the National Institute for Health and Clinical Excellence (NICE) in the UK, QALYs should be preferred as a measure of health effects over other measures such as life-years gained (Sanders et al. 2016; Whitehead and Ali 2010). Compared to other health outcomes, QALYs capture more comprehensively the impact of health interventions designed to improve both individual patient care and population-based health outcomes. They account not only for the length of life earned by adopting a specific intervention (i.e. life-years gained), but also for adverse events, and patients' experiences and preferences (i.e. health utilities). In fact, health utilities are based on preferences for the different health states that patients might experience, in such a way that the more desirable health states receive greater weight (Whitehead and Ali 2010). Moreover, QALYs provide a common measure for recording health effects, enabling comparisons across different areas of healthcare (Sanders et al. 2016). Thus, cost-utility analysis is the recommended methodology for evaluating the value of a medical intervention. Throughout this subchapter, the term cost-effectiveness analysis is used broadly and also refers to cost-utility analysis.

Components of the *value* beyond healthcare, which can have a considerable role in influencing decisions on health—such as ethical, legal and cultural issues, organisational aspects and consumer viewpoint—are not accounted for in cost-effectiveness analysis. This is particularly noteworthy in genomics, where social implications and context-related aspects play a central role in defining the overall *value* of a genomic application.

This subchapter describes the role of cost-effectiveness analysis in assessing genomic applications in PH and preventive medicine, and summarises the findings supporting the recommendations for implementing programmes for *BRCA1/2*-related cancers, Lynch syndrome (LS) and familial hypercholesterolaemia (FH).

Cost-Effectiveness Analysis and Health Programmes Involving Genomic Applications

Genomic applications in healthcare are very diverse, ranging from predictive, carrier and diagnostic genetic tests for rare Mendelian inherited diseases to compound genomic classifiers or biomarkers for the diagnosis and prevention of complex disorders or that inform drug therapies. Regardless of the technology in any particular case, a common first step is to provide a comprehensive definition of the genomic application, also called a *genomic health programme*. Four key elements must be fully described: performance characteristics of the technology, target population (e.g. newborn, general population, high-risk individuals), purpose of the intervention (e.g. screening, diagnosis or treatment), and pre- and/or post-test health interventions (Pitini et al. 2019; Teutsch et al. 2009).

The performance characteristics of the technology (e.g. test, classifier), such as analytic or clinical sensitivity and specificity, robustness, accuracy, positive and negative predictive values (see 3.1 "The HTA approach for the evaluation of genetic/genomic applications"), are not sufficient to define a genomic application, since these characteristics can vary depending on the intended use of the application. Thus, the other key elements mentioned above must be considered. Health interventions include both preliminary tests (e.g. tests for microsatellite instability or immunohistochemistry before mismatch repair testing for LS) or pre-test counselling (e.g. family history in BRCA-related cancers or familial hypercholesterolaemia), and post-test clinical management based on the results of the genomic testing (e.g.

routine follow-up examinations, surgical intervention, drug therapies, predictive testing available to family members) (Pitini et al. 2019; Teutsch et al. 2009). Accordingly, the unit of analysis of economic evaluation in genomics is not limited to the technology itself but the wider *genomic health programme* defined by the four key elements listed above. Therefore, cost-effectiveness analysis in genomics typically compares alternative genomic health programmes in terms of costs and health benefits (D'Andrea et al. 2015).

The Dual Purpose of Cost-Effectiveness Analysis in Genomics

After the development of a genomic application, evidence of its efficacy and effectiveness (i.e. evidence of analytic validity, clinical validity, clinical and/or personal utility) should be provided to support its use in clinical practice (Col 2003; Teutsch et al. 2009). Assuming this clinical research step is positive, i.e. analytic validity, clinical validity and clinical utility have been demonstrated, an economic evaluation, which represents the first step of the dissemination and implementation research process, should follow (D'Andrea et al. 2018; Marzuillo et al. 2014a). Cost-effectiveness analysis serves a dual purpose for policy makers: (i) to identify all feasible health programmes involving a defined use of a particular genomic application, and (ii) to recognise which health programmes can maximise the *value* of that genomic application.

The first requirement for any cost-effectiveness analysis is that the description of the alternatives being compared is detailed and comprehensive. As far as possible, all genomic health programmes that answer the same research/study question should be reviewed and included. However, this entails a multidisciplinary approach, especially when a genomic application has never been adopted in clinical practice. Collecting the necessary data on genomic health programmes offers particular challenges, because the use of genomic applications involves the experience and knowledge of many stakeholders (e.g. patients, clinicians—including geneticists and genetic consultants, decision analysts or health economists, ethicists, lawyers, psychologists, healthcare system policy makers, funding bodies, and PH officials, etc.). However, in practice, it is unlikely that all the necessary resources and information will be available so that all potentially comparable programmes can be investigated in a single economic evaluation, especially before a genomic application enters the market. This is a major concern, which the scientific community has tried to overcome by improving the comparability across studies through the so-called *reference case*.

The basic concept of the *reference case* is that findings from different cost-effectiveness analyses can be complementary and, even if they are not outcomes of the same economic analysis, they are still comparable and can be integrated. The *reference case* is "a set of standard methodological practices that all cost-effectiveness analyses should follow" to increase comparability and quality, and to limit the waste of time and resources resulting from the unnecessary replication of studies (Sanders et al. 2016). When a genomic application has been used in clinical practice for sufficient time (for example, the well-established *BRCA1/2* or mismatch repair gene tests), systematic reviews of economic evaluations might help to collect and define all genomic health programmes formally analysed over the years (D'Andrea et al. 2016; Di Marco et al. 2018).

The results of a cost-effectiveness study should also give definitive information on the *value* of a genomic health programme through analysis of incremental cost-effectiveness ratios. In this conventional measure, differences in health effects between two alternatives are reported in the denominator, while differences in costs are reported in the numerator. Reporting of net monetary benefit, net health benefit and cost-effectiveness thresholds is also encouraged. The use of a standardised method to report results is also one of the recommendations for the *reference case*. Other recommendations aimed at increasing comparability across studies are: provide results from the *reference case* based on both the healthcare sector perspective and the societal perspective; measure health effects in terms of QALYs; ensure transparency about all steps in the analysis; use sensitivity analysis to test assumptions and perspectives (Sanders et al. 2016).

When Do We Perform a Cost-Effectiveness Analysis?

Cost-effectiveness analysis might assist the dissemination and implementation process in at least three ways (Fig. 3.1).

Economic evaluation can be conceived as an *external* or *third-party* assessment where authoritative institutions and organisations—with no involvement in the development of the application, nor in the final decisions on implementation—oversee the production of economic reports that provide decision makers with objective information on the economic *value* of the genomic application (Fig. 3.1a, b). This *third-party* assessment can be integrated in the flow of the evaluation process with two different approaches.

In the first approach (Fig. 3.1a), cost-effectiveness analysis is one of the tools reviewed to produce practice guidelines, as well as clinical trials, non-randomised studies, systematic reviews and meta-analysis. Findings on the cost-effectiveness of *genomic health programmes* are integrated with those on the validity and utility of the genomic application used in the programmes. A practical example of this approach is given by the NICE in the UK, where economic evaluations are part of a structured process aimed at generating practice guidelines that can be adopted nationwide. Decision analysts and health economists form part of the core group responsible for the development of national clinical guidelines and have key roles in reviewing existing economic evaluations and/or conducting new economic analyses (NICE 2012). The main advantages of including data on cost-effectiveness in national practice guidelines, alongside analytical and clinical evidence, are that it reiterates the importance of the economic evidence as a significant part of the evaluation, dissemination and implementation process; it standardises the interpretation

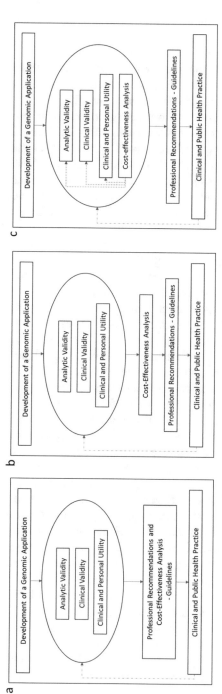

Fig. 3.1 Role of cost-effectiveness analysis of genomic applications. The flow charts show the structure of the evaluation process for a genomic application. Flow chart (**a**) integrates the performance of cost-effectiveness analysis with the development of practice guidelines. In flow chart (**b**) the cost-effectiveness analysis is not strictly related to the development of clinical guidelines and it will mainly be used by healthcare system policy makers as an instrument for price negotiation or for supporting the implementation of the test. Flow chart (**c**) integrates the performance of cost-effectiveness analysis in the early stages of the evaluation process to assess which, if any, characteristics of a genomic application need to be improved. The solid and dashed lines indicate the direction of the process

of the cost-effectiveness results; and it guarantees to all stakeholders—not only healthcare system policy makers and funding bodies, but also patients, clinicians and public health officials—easy access to the most comprehensive summary of the available information on the economic value of a specific genomic application.

In the second approach (Fig. 3.1b), economic evidence is produced independently of the development of practice guidelines. Economic evaluation is considered a more direct tool for price negotiation and reimbursement. Sensitivity analyses can be used to estimate a price range within which a new genomic health programme is cost-effective compared to others or in reference to a specific threshold (a threshold represents an estimate of how much healthcare systems or funding bodies are willing to pay for the health benefit, usually measured as one Life Years Gained (LYG)) (Bertram et al. 2016). For example, a population-based screening programme that offered *BRCA1/2* genetic tests to women without cancer was not found to be cost-effective compared to other programmes that offer the same genetic tests to selected women only. Moreover, the economic value of this population-based programme exceeded a reasonable threshold by a significant amount, showing that the programme was not good value for money (D'Andrea et al. 2016). On the other side, a programme that offers a test to all women might prevent more cancers than a programme that includes only part of a population, especially if the tool used to select the target population is not perfect. This might be a concern for *BRCA*-related cancers because the survey used to select the at-risk women focuses on ancestry, personal and familial history, and it might not capture new cases of "de novo" *BRCA* pathogenic variants occurring in individuals without personal or familial history. However, the high cost of the programme, to which the cost of *BRCA1/2* genetic testing itself makes a significant contribution, is still a major barrier to its implementation. A cost-effectiveness analysis might be useful for estimating at what *BRCA1/2* testing price range the population-based programme becomes a reasonable option in terms of costs and benefits. The results from the analysis might subsequently be used by decision makers to negotiate the costs of genetic testing with the supplier. Note that this example is only indicative and does not consider important implications of population-based screening in genomics that impact psychological, ethical, legal and social issues, as well as other indirect costs. A practical case of an independent organisation of this approach is given by the Institute for Clinical and Economic Review (ICER). Even though the economic reports produced by the ICER are not part of a structured process for reimbursement or approval of drugs and medical devices, they have a growing impact on price negotiations and formulary management (Neumann et al. 2018).

The third approach (Fig. 3.1c) is to conduct an economic evaluation early in the evaluation process to assess what characteristics of a genomic application need to be improved so that the corresponding genomic health programmes become cost-effective (e.g. how accurate a genotyping test needs to be, how large a population needs to be targeted). This approach can also be conceived as an *internal* process, since it can be conducted by the suppliers to improve the analytical or clinical characteristics of their final product and to better define their target population. In recent years, many genomic companies have funded studies not only to produce evidence on the effectiveness of their genomic devices, but also on their cost-effectiveness; in this way, they aim to better understand the potential benefits and limitations of their applications.

The three approaches described above are not mutually exclusive and can co-exist in the same system. However, it is imperative not to replicate efforts and, where possible, stakeholders and decision makers have to cooperate, so that the same economic evaluation can be used for different purposes, e.g. integrating the evidence of cost-effectiveness into clinical guidelines, setting the price of a new genomic application, or improving the *value* of a new technology.

Finally, it is important to emphasise that the process shown in Fig. 3.1 is dynamic. The analytic validity, clinical validity and clinical utility of a genomic application should be reconsidered after it is implemented in clinical practice, especially when changes occur in a genomic health programme (e.g. a new improved genomic technology is introduced, the costs of an application are reduced, the target population increases or decreases, alternative pre- and post-test interventions become available). For example, after two decades of using *BRCA1/2*-only testing for detecting BRCA-related cancers, a multiple-gene sequencing approach was recently introduced in clinical practice to assess up to 90 cancer-associated genes, including the *BRCA1/2* genes (Kurian et al. 2018). The health benefits and costs of adopting this new technology in genomic health programmes for *BRCA*-related cancers need to be explored, taking into account any additional costs and implications for other types of cancer.

In conclusion, cost-effectiveness analysis should be formally integrated into the evaluation process for any genomic health programme, where it should function as a bridge between clinical research and dissemination and implementation research.

Case Studies of Cost-Effective Genomic Health Programmes Ready to be Implemented in Clinical and Public Health Practice

Relatively few genomic applications are formally evaluated for their cost-effectiveness before entering clinical practice, and most of them have no evidence of clinical utility and clinical validity (D'Andrea et al. 2015; Phillips et al. 2014) (Fig. 3.1). Moreover, those with a potentially remarkable impact on public health at acceptable cost are underutilised (CDC 2019). In the following, we briefly summarise three paradigmatic examples of genomic health programmes with evidence of analytic validity, clinical validity, clinical utility and high, cost-effective *value* that are nevertheless not yet uniformly implemented in developed countries. The identification of the cost-effective programmes in this section was performed though systematic reviews of available economic evidence.

Genomic Health Programmes for *BRCA*-Related Cancers

Inherited mutations in breast/ovarian cancer susceptibility genes (*BRCA1, BRCA2*) are responsible for 2–7% of breast cancers and 10–15% of ovarian cancers (D'Andrea et al. 2016). Implementation of genomic health programmes involving the use of *BRCA* genetic testing would reduce the burden of *BRCA*-related cancers. Currently, there is evidence that four categories of *BRCA1/2* screening programme are good value for money from a healthcare perspective:

(i) *Population-based genomic screening based on ancestry*. Specifically, *BRCA1/2* screening of Ashkenazi Jewish populations because of the higher risk of specific *BRCA* gene mutations compared to the general population;

(ii) *Family history-based genomic screening* (i.e. testing individuals without cancer but with family history suggestive of *BRCA* mutation). Administration of a *BRCA* risk assessment tool, which can involve primary-care physicians, to detect at-risk individuals in possible *BRCA* families;

(iii) *Familial mutation-based genomic screening* (i.e. testing individuals without cancer, but who are related to an individual with a known familial *BRCA* mutation, also called *cascade screening* of the relatives);

(iv) *Cancer-based genomic screening* (i.e. testing individuals with *BRCA*-related cancers followed by cascade testing of relatives). There is no evidence of cost-effectiveness for *BRCA1/2* screening of all newly diagnosed cases of breast cancer, but the cost-effectiveness of screening all new cases of ovarian cancers is currently under evaluation. However, screening programmes for *BRCA1/2* that include tools for identifying women who are likely to have an inherited form of breast or ovarian cancer is cost-effective when followed by *BRCA1/2* testing of the carriers' relatives.

The design of the screening programme and the subsequent implementation process need to be tailored to the characteristics of the respective target population and healthcare system to ensure the translation of cost-effectiveness evidence into practice. The cost of *BRCA1/2* testing was of paramount importance in determining the cost-effectiveness of these programmes (D'Andrea et al. 2016).

Genomic Health Programmes for Lynch Syndrome

LS accounts for about 3% of all newly diagnosed colorectal cancer (CRC) cases, and it is caused by a mutation in one of the DNA mismatch repair genes (*MLH1, MSH2, MSH6* and *PMS2*) (Di Marco et al. 2018). LS is also associated with an increased risk of other cancers, such as endometrial cancer. From a healthcare perspective, there are two categories of screening programme that represent good value for money:

(i) Universal CRC-based LS screening (i.e. screening of newly diagnosed CRC patients, without performing a preliminary selection in terms of age or clinical criteria);

(ii) Screening of only those newly diagnosed CRC patients who fall below an age cut-off of 70 years.

As before, for efficient research translation, the genomic health programmes should be tailored to the respective target populations and healthcare systems (Di Marco et al. 2019; Pastorino et al. 2020; Tognetto et al. 2019).

Genomic Health Programmes for Familial Hypercholesterolaemia

FH is a monogenic disorder of lipid metabolism that leads to premature coronary artery disease and affects approximately 14 to 34 million individuals worldwide (Rosso et al. 2016). The following is a recommendation of demonstrated cost-effectiveness:

(i) Cascade screening to detect heterozygous pathogenic variants in one of three genes (*APOB*, *LDLR*, *PCSK9*) should be conducted on relatives of individuals with FH, including first- and second- and, where possible, third-degree biological relatives (Rosso et al. 2017b; Migliara et al. 2017).

The genetic testing can be combined with measurements of LDL-C concentration for clinical diagnosis. As recommended by current guidelines, in the absence of a DNA diagnosis, cascade testing using LDL-C concentration measurements should be undertaken to identify people with FH (NICE 2019).

Assessing the *value* of genomic healthcare programmes is not confined to genetic testing. The availability of pharmacogenomic treatments, genomic panel tests and classifiers is constantly growing (D'Andrea et al. 2019; D'Andrea et al. 2020) and these applications must undergo a similar evaluation process before entering clinical practice.

3.3   Genomics Training Needs of Public Health Professionals

Annalisa Rosso

Genomic Science and Public Health

With the completion of the first draft of the human genome sequence in 2003, and the numerous findings on the association between genes and disease susceptibility, there has been a debate on the utility of genomic science in PH. Human genome discoveries have led to a new approach to health and disease management, known as personalised medicine (PM), where all the various stages in the process, from prevention to diagnosis and treatment, may be customised for each individual based on genetic susceptibility (Boccia 2014). However, this approach seems to contradict the mission of PH, which is to improve health from a population perspective, and the incorporation of genomics into PH must therefore deal with this paradox (Khoury et al. 2011).

Some PH professionals have considered a limited role for genomics in PH, which should instead focus on the environmental causes of disease (i.e. infectious, chemical, behavioural and social factors) (Chowkwanyun et al. 2018). It has also been argued that investing in genomics may divert resources away from delivering basic PH services and addressing social and environmental determinants of disease and that most PH practitioners do not have the knowledge to integrate rapidly emerging genomic information into their programmes (Khoury et al. 2011). On the other hand, PH professionals and associations worldwide have stressed that the potential of PM to identify high-risk individuals and develop tailored preventive interventions cannot be ignored (Horton 2018).

Over recent decades, several practitioners have supported the incorporation of genome-based knowledge and technologies into PH, leading to the emergence of PHG as a multidisciplinary field that establishes scientific and policy foundations for the appropriate translation of genomics research into health benefits for individuals and populations. It has been suggested that PH professionals may play different roles in this translation process. First, they may act as "honest brokers", responsible for evaluating the effectiveness of genomic applications, who prevent potential harm and unnecessary healthcare expenditure through premature use (Khoury et al. 2011). In fact, the rapidly increasing number of genomic applications requires a continuously updated, transparent assessment of their effectiveness and cost-effectiveness to avoid unnecessary healthcare expenditure on unwarranted tests, to minimise harm deriving from the unintended consequences of premature use of technology, and to maximise the potential population health benefits. Furthermore, given the cost constraints currently faced by publicly funded healthcare systems, PH professionals are likely to support the prioritisation of genetic services that should be funded from public budgets, based on criteria of clinical utility and validity, but also on organisational aspects, economic evaluation, ELSI and patient perspective (see 3.1 "The HTA approach for the evaluation of genetic/genomic applications").

PH professionals may also use genomics tools to plan and evaluate the health impact of PH interventions, including those that do not involve genetics, on different subsets of the population. For example, genetic information, such as that easily collected through family history, could be used to identify subsets of the population at higher risk of developing a disease, to which more intensive or different types of intervention should be addressed. Genomic information could also be used as a stratifying tool for measuring the health impact of interventions on different subsets of the population, based on their individual susceptibility (Khoury et al. 2011). In this regard, the concept of "precision PH" has recently been developed, which is defined as "the application and combination of new and existing technologies, which more precisely describe and analyse individuals and their environment over the life course, in order to tailor preventive interventions for at-risk groups and improve the overall health of a population" (Weeramanthri et al. 2018). Building upon the work conducted in the PHG field during the last 20 years, precision PH could therefore enable the development of more precise individual and population-based interventions, contributing at the same time to an improvement in population health outcomes, thus overcoming the dichotomy of high-risk versus population-based approaches (Molster et al. 2018).

Most importantly, PH professionals can contribute to modelling and evaluating the implementation of evidence-based genomic applications at multiple levels (Khoury et al. 2011), which would support the appropriate integration of genomic knowledge and technologies into all aspects of PH. Depending on the specific clinical condition under consideration, the implementation could include, for instance, the development of education and health-promotion campaigns targeted at at-risk individuals, or the integration of tools into existing PH programmes, such as cancer screening.

Genomics, Public Health Functions, Public Health Services

Building on these concepts, in the United States, genomics has been increasingly integrated into the three recognised core PH functions, i.e. assessment, policy development and assurance, and into the ten essential PH services defined by the Core Public Health Functions Steering Committee (CDC 2018). Table 3.2 (adapted from Molster et al. 2018) provides a synthesis of how genomics can be incorporated into PH practice, based on examples of regulations for laboratories using genome-based technologies emerging from published literature.

Thus, there is already solid evidence that genomics can contribute to PH practice in multiple ways. However, incorporating genomic advances into the delivery of essential PH services requires adequate competencies from the PH workforce. In this regard, since the early development of the PHG movement, ensuring an adequate PH capacity in genomics has been considered a priority (Khoury and Mensah 2005).

Competence of Public Health Professionals in Genomics

What is the current level of competence of PH professionals in genomics? While several surveys have been performed to evaluate the knowledge, attitudes and professional behaviour of physicians towards the integration of human genomic discoveries into clinical practice (e.g. Marzuillo et al. 2013; Petersen et al. 2014), there is little evidence on the capacity of PH practitioners in this area.

The first study that aimed to assess the attitudes and knowledge of PH professionals in genomics surveyed PH educators across the United States (Chen and Goodson 2007). In particular, these authors wanted to understand the readiness of

Table 3.2 Ten essential public health services and related PHG activities (adapted from Molster et al. 2018)

Essential public health services	PH activities
1. Monitor health status to identify and solve community health problems	• Assess the distribution and impact of genetic risk factors to determine their contribution to health status and the burden of disease • Promote the development of resources that enable monitoring of the genomic-related health status of populations
2. Diagnose and investigate health problems and health hazards in the community	• Identify and track infectious disease outbreaks using genomic technology • Assist with the redesign of diagnostic and laboratory services to incorporate new genome-based technologies
3. Inform, educate, and empower people about health issues	• Improve the genomic literacy of the public • Empower all stakeholders, including health professionals and the public, to make informed decisions about the uses of genetic information with realistic expectations about the risks and benefits • Facilitate the integration of genomics into health promotion and disease prevention programs
4. Mobilise community partnerships and action to identify and solve health problems	Foster collaborations between stakeholders
5. Develop policies and plans that support individual and community health efforts	Policies and plans that could be developed include those relating to: • The appropriate use of genomic applications • Equity and accessibility • The use of family health history information • Reproductive decision-making
6. Enforce laws and regulations that protect health and ensure safety	Contribute to: • Laws and regulations for genomic applications; • Regulations for laboratories using genome-based technologies
7. Link people to needed personal health services and assure the provision of healthcare when otherwise unavailable	Support the appropriate integration of genomic knowledge and technologies into all aspects of healthcare and public health
8. Assure a competent public and personal healthcare workforce	• Contribute to training and education in, and development of, genomic knowledge, skills and capacity for health professionals • Support the development of workforce capacity in genomics-related fields
9. Evaluate effectiveness, accessibility, and quality of personal and population-based health services	• Evaluate new genome-based knowledge and technologies to determine their evidence base, quality, appropriateness and readiness for implementation in healthcare and public health practice • Evaluate the use of genome-based knowledge and technologies in healthcare and public health practice

(continued)

Table 3.2 (continued)

Essential public health services	PH activities
10. Research new insights into, and innovative solutions for, health problems	• Monitor the results of human genome epidemiology studies • Support the development of infrastructure to conduct genome-related population research • Conduct and monitor translation research

Source: Adapted from Molster CM, Bowman FL, Bilkey GA, et al. (Molster et al. 2018) The Evolution of Public Health Genomics: Exploring Its Past, Present, and Future. Front. Public Health 6:247, Table 1. Public health genomics activities in relation to the 10 essential public health services. Some modifications to the text were made. https://doi.org/10.3389/fpubh.2018.00247, licensed under the terms of the Creative Commons Attribution License (https://creativecommons.org/licenses/by/4.0/)

PH educators to incorporate into their work the genomic competencies developed by the CDC for public health educators "as a tool for PH programmes and schools of PH to incorporate genomics into existing competencies and programme training goals" (CDC 2001). Most of the 1607 PH educators who responded to the survey either agreed or strongly agreed with genomic competencies being proposed for the PH education workforce. However, less than half the sample (49.3%) considered it important in PH education. A section of the survey was designed to assess the respondents' knowledge of basic and applied genomics, and found this to be deficient in the sample: on average, 51.1% of the answers to six items on genomics were correct (equivalent to a university "F" grade). Almost half the respondents stated they could not make appropriate PH recommendations based on their clients' family histories. The study also aimed to assess the determinants of attitudes towards genomic education competencies, finding a significant association between positive attitudes and genomic knowledge, even after controlling for socio-economic factors, awareness of PH education policies, and training in genomics. Thus, the authors concluded that relevant training in genomics for this group of health professionals should be developed and promoted, and that continuing education tools, focusing on PH genomics content, might be a means of delivering information and developing favourable professional attitudes.

In a subsequent analysis conducted on the same dataset, Chen et al. examined the likelihood that US health educators would incorporate genomic competencies into their health education programmes, and the factors influencing such likelihood, through the development and testing of a conceptual, theory-based model (Chen and Goodson 2008). The authors reported a low likelihood that US health educators would adopt genomic competencies (29.3% in average). How the participants perceived the compatibility between PHG and their professional and personal roles, their perceptions of the complexity of genomics, and the communication channels used to learn about PHG, significantly affected their genomic knowledge and attitudes. Given the negative attitudes found in the sample, the authors stressed the need to conduct future research on the topic and to train the PH workforce in genomics. A subsequent mixed-methods analysis they conducted identified as barriers to adopting genomics into public health education the lack of genomic knowledge (basic and applied), having to deal with the lay public's reaction, lack of priority, time and resources, and incompatibility between genomics and public health educators' religious and ethical beliefs (Chen and Goodson 2009). Despite the importance of these conclusions, no other studies aimed at assessing the competencies and attitudes of the PH workforce in genomics have been published in the United States since the work of Chen et al.

Some years later, the Department of Public Health and Infectious Diseases of Sapienza University of Rome conducted the first survey to specifically assess the knowledge, attitudes and training of PH professionals in the field of predictive genetic testing for chronic diseases (Marzuillo et al. 2014b). The survey involved a sample of PH professionals randomly selected from the register of the Italian Society of Hygiene, Preventive Medicine and Public Health (S.It.I.). The study showed that Italian PH professionals had the necessary attitudinal background to contribute to the proper use of predictive genetic testing for chronic diseases, but they needed additional training to increase their methodological knowledge. Only 10.2% of the respondents answered all seven questions about predictive genetic testing correctly and this knowledge was significantly associated with exposure to predictive genetic testing during postgraduate training and with time dedicated to continuing medical education. About 80% of PH professionals considered their knowledge inadequate and 86.0% believed that it should be improved through specific postgraduate training courses. Adequate knowledge was the strongest predictor of positive attitudes towards genomics from a PH perspective.

In 2016/2017, the same department conducted a similar survey of a sample of European PH professionals with membership of the European PH Association (EUPHA), which represents European PH professionals, to assess their attitudes towards their role in the implementation of PHG, and their knowledge and attitudes regarding genetic testing and genetic services (Rosso et al. 2017a). This survey found that knowledge of genomics among PH professionals was also low at the European level: for instance, only 28.9% of the respondents correctly identified all medical conditions for which there is (or is not) evidence that supports the implementation of genetic testing. On the other hand, in general, attitudes towards both the use of genetic testing and delivery of genetic services, and the involvement of PH professionals in putting PHG into practice, were positive. The only negative sentiment was in response to the proposal that it was more important to invest resources in the social and environmental causes of ill health than in the implementation of genetic testing. Thus, over 60% of respondents agreed that PH resources should mainly be targeted at addressing the structural causes of ill health. Similarly to Chen et al. and Marzuillo et al., Rosso et al. also found that better knowledge was associated with positive attitudes towards the use of genetic testing and the delivery of genetic services in PH. There was strong agreement that PH professionals should foster the integration of genomics into PH practices, and over 80% of respondents agreed that, in future, PH programmes will make greater use of genetic information.

Positive attitudes towards the involvement of PH professionals in genomics were associated with being trained in genetic testing during undergraduate education and with direct involvement in PHG activities.

In conclusion, the limited available evidence on the knowledge and attitudes of PH professionals relating to genomics shows overall positive attitudes towards the incorporation of genomics in PH practice, but controversial results on the perceived value of genetic technologies for PH purposes. Positive attitudes are associated with better knowledge and training in genomics. In this regard, a more widespread awareness of those genetic and genomic applications with documented evidence of effectiveness and cost-effectiveness could probably improve attitudes towards the inclusion of genomics in preventive interventions. Given the potential contribution of genomics in essential public health services, as outlined in Table 3.2, efforts should be made to increase the capacity of the PH workforce in the three main domains of assessment, policy development and implementation and evaluation of genomic technologies.

References

Battista, R. N. (2006). Expanding the scientific basis of health technology assessment: A research agenda for the next decade. *International Journal of Technology Assessment in Health Care, 22*(3), 275–280.

Bertram, M. Y., Lauer, J. A., De Joncheere, K., et al. (2016). Cost-effectiveness thresholds: Pros and cons. *Bulletin of the World Health Organization, 94*(12), 925–930.

Boccia, S. (2014). Why is personalized medicine relevant to public health? *European Journal of Public Health, 24*, 349–350.

Boccia, S., Federici, A., Colotto, M., et al. (2014). Implementation of Italian guidelines on public health genomics in Italy: A challenging policy of the NHS. *Epidemiologia e Prevenzione, 38*(6 Suppl 2), 29–34.

Burke, W., & Zimmern, R. (2007). *Moving beyond ACCE: An Expanded Framework for Genetic Test Evaluation. A paper for the United Kingdom Genetic Testing Network*. Cambridge, UK: PHG Foundation.

Burke, W., Atkins, D., Gwinn, M., et al. (2002). Genetic test evaluation: information needs of clinicians, policy makers, and the public. *American Journal of Epidemiology, 156*(4), 311–318.

Centers for Disease Control and Prevention – CDC. (2001). Genomic competencies for the public health workforce. Retrieved from https://www.cdc.gov/genomics/translation/competencies/index.htm#education. Accessed 4 Dec 2019.

Centers for Disease Control and Prevention. (2010). ACCE Model List of 44 Targeted Questions. CDC. Retrieved from https://www.cdc.gov/genomics/gtesting/acce/acce_proj.htm. Accessed 4 Dec 2019.

Centers for Disease Control and Prevention – CDC. (2018). Retrieved from https://www.cdc.gov/publichealthgateway/publichealthservices/essentialhealthservices.html. Accessed 4 Dec 2019.

Centers for Disease Control and Prevention – CDC. (2019). Genomics & Precision Health. Tier 1 Genomics Applications and their Importance to Public Health. Retrieved from https://www.cdc.gov/genomics/implementation/toolkit/tier1.htm. Accessed 3 Dec 2019.

Chen, L. S., & Goodson, P. (2007). Public health genomics knowledge and attitudes: A survey of public health educators in the United States. *Genetics in Medicine, 9*, 496–503.

Chen, L. S., & Goodson, P. (2008). US health educators' likelihood of adopting genomic competencies into health promotion. *American Journal of Public Health, 98*, 1651–1167.

Chen, L. S., & Goodson, P. (2009). Barriers to adopting genomics into public health education: A mixed methods study. *Genetics in Medicine, 11*, 104–110.

Chowkwanyun, M., Bayer, R., & Galea, S. (2018). "Precision" PH - Between novelty and hype. *The New England Journal of Medicine, 379*, 1398–1400.

Col, N. F. (2003). The use of gene tests to detect hereditary predisposition to chronic disease: Is cost-effectiveness analysis relevant? *Medical Decision Making, 23*(5), 441–448.

D'Andrea, E., Marzuillo, C., Pelone, F., et al. (2015). Genetic testing and economic evaluations: A systematic review of the literature. *Epidemiologia e Prevenzione, 39*(4 Suppl 1), 45–50.

D'Andrea, E., Marzuillo, C., De Vito, C., et al. (2016). Which BRCA genetic testing programs are ready for implementation in health care? A systematic review of economic evaluations. *Genet Med, 18*(12), 1171–1180.

D'Andrea, E., Lagerberg, T., De Vito, C., et al. (2018). Patient experience and utility of genetic information: A cross-sectional study among patients tested for cancer susceptibility and thrombophilia. *European Journal of Human Genetics, 26*(4), 518–526.

D'Andrea, E., Ahnen, D. J., Sussman, D. A., & Najafzadeh, M. (2019). Quantifying the impact of adherence to screening strategies on colorectal cancer incidence and mortality. *Cancer Medicine, 9*, 824–836. https://doi.org/10.1002/cam4.2735.

D'Andrea, E., Choudhry, N. K., Raby, B., et al. (2020). A bronchial-airway gene-expression classifier to improve the diagnosis of lung cancer: Clinical outcomes and cost-effectiveness analysis. *International Journal of Cancer, 146*(3), 781–790.

Di Marco, M., D'Andrea, E., Panic, N., et al. (2018). Which Lynch syndrome screening programs could be implemented in the "real world"? A systematic review of economic evaluations. *Genetics in Medicine, 20*(10), 1131–1144.

Di Marco, M., D'Andrea, E., Villari, P., et al. (2019). Universal screening of Lynch syndrome is ready for implementation. *Genetics in Medicine, 21*(1), 254–255.

EUnetHTA Joint Action 2 WP 8. (2013). HTA Core Model ® version 3.0. Retrieved from http://eunethta.eu/sites/5026.fedimbo.belgium.be/files/HTACoreModel3.0.pdf. Accessed 4 Dec 2019.

Giacomini, M., Miller, F., & Browman, G. (2003). Confronting the "Gray Zones" of Technology Assessment: evaluating genetic testing services for public insurance coverage in Canada. *International Journal of Technology Assessment in Health Care, 19*(2), 301–316.

Horton, R. (2018). Offline: In defence of precision public health. *Lancet, 392*, 504.

Khoury, M. J. (2017). No Shortcuts on the Long Road to Evidence-Based Genomic Medicine. *JAMA, 318*(1), 27.

Khoury, M. J., & Mensah, G. A. (2005). Genomics and the prevention and control of common chronic diseases: emerging priorities for public health action. *Prev Chronic, 2*, A05.

Khoury, M. J., Bowen, M. S., Burke, W., et al. (2011). Current priorities for PH practice in addressing the role of human genomics in improving population health. *American Journal of Preventive Medicine, 40*, 486–493.

Kohler, J. N., Turbitt, E., & Biesecker, B. B. (2017). Personal utility in genomic testing: A systematic literature review. *European Journal of Human Genetics, 25*(6), 662–668.

Kurian, A. W., Ward, K. C., Hamilton, A. S., et al. (2018). Uptake, results, and outcomes of germline multiple-gene sequencing after diagnosis of breast cancer. *JAMA Oncology, 4*(8), 1066–1072.

Marzuillo, C., De Vito, C., Boccia, S., et al. (2013). Knowledge, attitudes and behavior of physicians regarding predictive genetic tests for breast and colorectal cancer. *Preventive Medicine, 57*(5), 477–482.

Marzuillo, C., De Vito, C., D'Addario, M., et al. (2014a). Are public health professionals prepared for public health genomics? A cross-sectional survey in Italy. *BMC Health Services Research, 28*(14), 239.

Marzuillo, C., De Vito, C., D'Andrea, E., et al. (2014b). Predictive genetic testing for complex diseases: A public health perspective. *QJM, 107*(2), 93–97.

Migliara, G., Baccolini, V., Rosso, A., et al. (2017). Familial hypercholesterolemia: A systematic review of guidelines on genetic testing and patient management. *Front Public Health, 5*, 252.

Molster, C. M., Bowman, F. L., Bilkey, G. A., Cho, A. S., Burns, B. L., Nowak, K. H., & Dawkins, H. J. (2018). The evolution of public health genomics: exploring its past, present, and future. *Frontiers in Public Health, 6*, 247.

National Institute for Health and Care Excellence (NICE). (2012). The guidelines manual. Process and methods [PMG6]. 7 Assessing cost effectiveness. 7.1 The role of the health economist in clinical guideline development. Retrieved from https://www.nice.org.uk/process/pmg6/chapter/assessing-cost-effectiveness#ftn.footnote_13. Accessed 3 Oct 2019.

National Institute for Health and Care Excellence (NICE). (2019). FH recommendation (NICE Clinical Guideline 71 – Identification and Management of Familial Hypercholesterolemia). Retrieved from https://www.nice.org.uk/guidance/qs41. Accessed 3 Oct 2019.

NEJM Catalyst. (2017). What is Patient-Centered Care? NEJM Catalyst. Retrieved from https://catalyst.nejm.org/what-is-patient-centered-care/. Accessed 4 Dec 2019.

Neumann, P. J., Rosen, A. B., & Weinstein, M. C. (2005). Medicare and cost-effectiveness analysis. *The New England Journal of Medicine, 353*(14), 1516–1522.

Neumann, P. J., Silver, M. C., Cohen, J. T., et al.. (2018). Should a drug's value depend on the disease or population it treats? Insights from ICER's value assessments. Health Affairs blog. Retrieved from https://www.healthaffairs.org/do/10.1377/hblog20181105.38350/full/. Accessed 3 Oct 2019.

Pastorino, R., Basile, M., Tognetto, A., et al. (2020). Cost-effectiveness analysis of genetic diagnostic strategies for Lynch syndrome in Italy. *PLoS One, 15*(7), e0235038.

Petersen, K. E., Prows, C. A., Martin, L. J., et al. (2014). Personalized medicine, availability, and group disparity: an inquiry into how physicians perceive and rate the elements and barriers of personalized medicine. *Public Health Genomics, 17*(4), 209–220.

Phillips, K. A., Ann Sakowski, J., Trosman, J., et al. (2014). The economic value of personalized medicine tests: What we know and what we need to know. *Genetics in Medicine, 16*(3), 251–257.

Pitini, E., De Vito, C., Marzuillo, C., et al. (2018). How is genetic testing evaluated? A systematic review of the literature. *European Journal of Human Genetics, 26*(5), 605–615.

Pitini, E., D'Andrea, E., De Vito, C., et al. (2019). A proposal of a new evaluation framework towards implementation of genetic tests. *PLoS One, 14*(8), e0219755.

Porter, M. E. (2010). What is value in health care? *The New England Journal of Medicine, 363*(26), 2477–2481.

Rosso, A., D'Andrea, E., Di Marco, M., et al. (2017a). Interim results of EUPHA network members' s survey on Public Health Genomics. *Eur J Public Health, 27*(suppl 3), ckx187.327.

Rosso, A., Pitini, E., D'Andrea, E., et al. (2017b). The cost-effectiveness of Genetic Screening for Familial Hypercholesterolemia: A Systematic. *Review, 29*(5), 464–480.

Sanders, G. D., Neumann, P. J., Basu, A., et al. (2016). Recommendations for conduct, methodological practices, and reporting of cost-effectiveness analyses: second panel on cost-effectiveness in health and medicine. *JAMA, 316*(10), 1093–1103.

Schmidtke, J., & Cassiman, J.-J. (2010). The EuroGentest clinical utility gene cards. *European Journal of Human Genetics, 9*, 1068.

Teutsch, S. M., Bradley, L. A., Palomaki, G. E., et al. (2009). The evaluation of genomic applications in practice and prevention (EGAPP) initiative: Methods of the EGAPP Working Group. *Genetics in Medicine, 11*(1), 3–14.

Tognetto, A., Pastorino, R., Castorina, S., et al. (2019). The current practice of lynch syndrome diagnosis and management in Italy: A qualitative assessment. *Public Health Genomics, 22*(5–6), 189–207.

Unim, B., Pitini, E., & Lagerberg, T. (2019). Current genetic service delivery models for the provision of genetic testing in Europe: A systematic review of the literature. *Front Genet, 10*, 552.

Weeramanthri, T. S., Dawkins, H. J. S., Baynam, G., et al. (2018). Editorial: Precision public health. *Frontiers in Public Health, 6*, 121.

Whitehead, S. J., & Ali, S. (2010). Health outcomes in economic evaluation: the QALY and utilities. *British Medical Bulletin, 96*, 5–21.

Chapter 4
Ethico-legal and Policy Issues Surrounding Personalised Medicine

Roberta Pastorino, Michael Lang, Ma'n H. Zawati, Giovanna Elisa Calabrò, Ilda Hoxhaj, Elisa J. F. Houwink, Michele Sassano, and Stefania Boccia

4.1 Ethical Issues in Personalised Medicine: A Survey and the Path Ahead

Michael Lang and Ma'n H. Zawati

Rapidly proliferating advances in genomic medicine have made it possible to more precisely target clinical treatment and preventative care to atomised groups or individual patients. Ongoing changes in healthcare delivery, called "personalised medi-

R. Pastorino
Department of Woman and Child Health and Public Health, Public Health Area, Fondazione Policlinico Universitario A. Gemelli IRCCS, Rome, Italy
e-mail: roberta.pastorino@policlinicogemelli.it

M. Lang · M. H. Zawati
McGill University Centre of Genomics and Policy, Montreal, QC, Canada
e-mail: michael.lang@mail.mcgill.ca; man.zawati@mcgill.ca

G. E. Calabrò · I. Hoxhaj (✉) · M. Sassano
Section of Hygiene, University Department of Life Sciences and Public Health, Università Cattolica del Sacro Cuore, Rome, Italy
e-mail: ilda.hoxhaj1@unicatt.it; michele.sassano02@icatt.it

E. J. F. Houwink
Department of Public Health and Primary Care (PHEG), Leiden University Medical Centre, Leiden, The Netherlands
e-mail: E.J.F.Houwink@lumc.nl

S. Boccia
Department of Woman and Child Health and Public Health, Public Health Area, Fondazione Policlinico Universitario A. Gemelli IRCCS, Rome, Italy

Section of Hygiene, University Department of Life Sciences and Public Health, Università Cattolica del Sacro Cuore, Rome, Italy
e-mail: stefania.boccia@unicatt.it; stefania.boccia@policlinicogemelli.it

S. Boccia et al. (eds.), *Personalised Health Care*, SpringerBriefs in Public Health, https://doi.org/10.1007/978-3-030-52399-2_4

57

cine (PM)" in the context of this essay, are revolutionising the management of clinical relationships and are contributing to improved patient outcomes. At the same time, the disruptions brought about by PM raise a number of pressing ethical challenges. Accounting for such challenges is essential for ensuring that PM is successfully implemented in treatment and prevention.

This essay will endeavour to outline some of the most prominent ethics issues in this context. We will predict how these issues are likely to affect the efficacy and efficiency of PM initiatives and will briefly propose how they may be mitigated. To begin, however, we will outline what the language of PM is typically understood to capture and how the choice to adopt such language might itself be a choice with certain ethical implications.

The Personalised Medicine Revolution

Early academic discussions on PM proposed a shift in the way that disease is conceived. On these accounts, PM is a departure from the dominant model of disease in the twentieth century, which identified disease according to tissue pathology and the presence of certain signs and symptoms (Whitcomb 2012). PM reflects a departure from this once dominant model, emphasising health instead of disease and pursuing optimised treatment and continuous optimisation (Whitcomb 2012). Following the completion of the Human Genome Project in 2003, stakeholders across the healthcare world have advocated for applying lessons from genomics to implement PM, thereby improving the "stratification and timing of healthcare by utilising biological information and biomarkers" (Schleidgen et al. 2013).

Several commentators argue that the concept of PM may be seen to exaggerate how finely tuned treatments targeted at individual patients are likely to be. For the foreseeable future, after all, PM is likely to be "more about categorising patients into different classes of generic risk and therapeutic efficacy" based on clinical genotyping than about designing individualised treatments (Juengst et al. 2016).

In the American context, the language of "precision medicine" has been promoted as an explanatory label that does not evoke the promise of individualised treatment but that also maintains a focus on molecular profiling as the basis for organising care (Juengst et al. 2016). These rhetorical debates, while not firmly the subject of this essay, may have important ethics implications. To the extent that personalised and precision medicine promise individually targeted treatments, for example, they may be seen to promise more than they are capable of delivering (Juengst et al. 2016). In what follows, we will primarily adopt the language of PM, recognising that the danger of overpromising outcomes may erode the public's trust in PM, potentially undermining its progress and hampering participant engagement (Adjekum et al. 2017). How such initiatives are brought to the public's attention is thus itself an important ethical issue.

Ethical Issues in Personalised Medicine

In this part of the essay, we will outline three of the most pressing ethical issues raised in the transition from twentieth century medicine to PM. The concerns we describe here are not, of course, the *only* ethical issues relevant in this space. Our primary objective is rather to survey the most pressing challenges and to consider how accounting for them will be vital in assuring the successful realisation of PM for prevention and treatment.

Research/Care Divide

Early discussions on PM emphasised its potential to reconceive of patients as partners in their care and in research initiatives leading to future health innovations (Collins and Varmus 2015; Blasimme and Vayena 2016; Hammer 2016). Increased connectivity, social media, and the use of mobile devices and smartphone applications with health-related functions all contribute to the sense that members of the public have a "growing desire to be active partners in medical research" (Collins and Varmus 2015). PM primarily facilitates this objective by encouraging patients to contribute health data and samples to largescale research initiatives both over the course of their treatment and in their lives outside of the clinic (Blasimme and Vayena 2016).

In a PM system, data collected in the clinic and subsequent refinements in treatment are "continuously fed back to improve [the] care of individual patients and contribute to the sum of medical knowledge" (Yuan 2019). This movement towards incorporating research into clinical care appears to be in clear conflict with the longstanding tendency to conceive of these activities as ethically distinct (Appelbaum et al. 2018). Considering that the primary objective of health research is to advance medical knowledge, while the primary objective of clinical care is to advance the best interest of the patient, the traditional view has been that these activities should be ethically distinguishable in practice (Appelbaum et al. 2018). In particular, ethicists have generally thought it is critical for participant consent to clearly specify that the interventions offered in the course of a research project have a primarily investigational purpose.

PM disrupts this narrative by collapsing medical care into research (Yuan 2019). By feeding clinical data into research initiatives that refine and modulate therapy, PM is difficult to categorise as uniquely fitting into research or care categories. Ethics discourse, which has been committed to a firm distinction between these concepts in recent decades, will thus likely need to evolve in the coming years. In the United States, for example, the Common Rule assumes a clear divide between research and care (Easter et al. 2006) and recent updates do not appear to have addressed the concern that this divide will impede novel forms of health research, including PM (Henry 2013). Consent processes should be adapted to communicate the particular ethical status of PM initiatives and to clarify their position between clinical care and research. As the transition towards PM quickens, it will be increasingly critical to consider the ethical implications of conceiving of patients as partners and conceiving of research and clinical care as unified enterprises.

Return of Results

PM, as we described above, is partially enabled by the availability of largescale genomic databases and novel computational tools for analysing complex datasets (Collins and Varmus 2015). One consequence of this widespread access to health data on the part of researchers and clinicians is the vastly increased probability that incidental findings will be discovered (Fiore and Goodman 2016). Such findings typically result from genome or exome sequencing and apply to matters unrelated to the issue for which the test was initially indicated (Green et al. 2013). In the context of a PM system in which genome sequencing is common, we can expect that numerous unintended findings about present or future condition susceptibility will be uncovered (Fiore and Goodman 2016).

What to do about such findings is a pressing and largely unsettled ethical and regulatory concern (Thorogood et al. 2019). For one thing, it is unclear whether clinicians and researchers have an ethical obligation to actively seek out potentially consequential findings beyond the ambit of a particular analysis (Schuol et al. 2015). By the contrary measure, it could be that researchers and clinicians have an ethical obligation to restrict their analysis to the specific set of questions in issue, that is, that they should systematically avoid uncovering unintended results (Fiore and Goodman 2016). Further complicating the issue, different policy guidelines suggest vastly different mechanisms for managing incidental findings. The American College of Medical Genetics and Genomics, for example, leaves the matter largely up to patients in the clinical context (David et al. 2019). A patient may only receive the secondary results arising from a genomic test if they initiate recontact with the relevant clinician for that purpose. On the approach taken by the American Society of Human Genetics, on the other hand, researchers are encouraged to initiate recontact with research participants to convey incidental findings or reinterpretations of results (Bombard et al. 2019). Against this backdrop, some scholars have pointed out that returning too much to patients and participants risks overwhelming them. At the same time, an onerous approach to returning results risks overburdening the researchers and clinicians, thereby potentially impeding research (Knoppers et al. 2019).

Underpinning this discussion, of course, is the promise that PM will endeavour to treat patients as partners in the health research and clinical care ecosystem. A critical element of such partnership is the return of benefit to the communities that have contributed data and samples to the PM system (Blasimme and Vayena 2016). The precise shape that such benefit will take, and how they will be returned to participating communities, is a pressing and unsettled ethical consideration.

Privacy

The massive volume of data upon which PM depends naturally leads to the creation of numerous novel privacy considerations. In particular, PM makes use of new and potentially sensitive forms of data to accomplish its objectives, including the results of whole genome sequencing and medical imaging. Such information is collected, stored, and shared with the research community, raising the possibility of privacy breach at nearly every stage. At the same time, PM promises to draw on sources of mobile health data, including information collected from individual participants via a mobile phone or other smart device. Such information, the "digital traces" of participants' daily lives, may have great utility for assessing human health and its relationship to social conduct and physical activity (Sankar and Parker 2017).

While such information is promising from a health research perspective, mobile health information may also reveal a great deal about a person's interests and lifestyle, thus heightening the privacy interests held in such data. Privacy protection mechanisms in PM should be attentive both to conventional forms of health data, such as test results and medical records, as well as other emerging and equally sensitive forms of information, such as that collected by smartphones in the course of daily life.

Broadly speaking, PM must work to balance two distinct interests: protecting participant information, including potentially sensitive genetic data, and assuring robust researcher access to information used to advance care (Sankar and Parker 2017). It will sometimes be challenging to advance both of these objectives at once. Measures taken to safeguard privacy, for example, can never achieve absolute success; the sharing of health information will always entail some risk that privacy will be compromised (Fiore and Goodman 2016). PM raises the ethical imperative to develop sufficiently strong privacy protections that are not so onerous as to stifle scientific advancement. To accomplish these aims, "reliable, comprehensive, state-of-the-art" information management systems must be paired with strong institutional policies that govern data access and use (Sankar and Parker 2017).

The Path Ahead

As PM initiatives become increasingly prominent fixtures in healthcare, the successful realisation of its objectives will depend in no small measure on the robust management of the ethical challenges it raises. Strong ethical principles will, for one thing, play a vital role in ensuring that the public can have confidence in PM. The implementation of PM, moreover, intimately depends on the existence of such confidence. Participants and patients need to be able to trust that their data will be used responsibly for the ultimate improvement of clinical care (Kraft et al. 2018). To this end, steps should be taken to better understand the ethical obligations of researchers, clinicians, patients, and the general public in the context of this rapidly shifting healthcare paradigm and how such obligations can be actualised in a manner that promotes public confidence and supports the expansion of knowledge about human health.

To be sure, the ethical issues examined above represent only some of those likely to be relevant in the coming decades. Issues surrounding patient autonomy, social identification, and genomics literacy, for example, will also be dominant concerns in future ethics deliberation. Such issues should be kept in consistent view. Our purpose here was rather to provide a brief introduction to the ethics of PM. In presenting this survey, we have also sought to highlight issues requiring closer attention. PM offers great promise, but its objectives are likely to be realised only insofar as the ongoing paradigm shift in treatment and care is carried out with watchful attention to its deeply ethical import.

4.2 Policy Issues of Genomics in Healthcare: A Focus on Training Needs for Healthcare Professionals

Giovanna Elisa Calabrò, Ilda Hoxhaj, Elisa J. F. Houwink, and Stefania Boccia

Policy Issues in Genomics

With the completion of sequencing the human genome, and hence the Human Genome Project, the traditional one-size-fits-all approach to disease management is expected to be progressively replaced by an individualised approach. This new model of personalised healthcare already had important implications for all the stakeholders, from individual citizens and physician practice to institutional, national, and international policymakers (Boccia 2012). The ability to respond to this change depends on a country capability and capacity to support implementation of genomics technologies into the national health system. In this sense, policymakers have a key role in implementing the relevant strategies for the integration of genomics into healthcare practice in the most effective possible way.

Engagement and support from institutional leadership is necessary to achieve transformational change in health systems and to facilitate the changes required. A challenge to the stakeholders remains the effective sustainable translation of the genomic knowledge into clinical practice. Messy translation of genomic advances into the clinic practice and unrealistic promises are limiting potential preventive treatments and improved medical care (Evans et al. 2011). An important role in facilitating the transformation of health systems during the implementation process is attributed to public health professionals. A widespread and coordinated collaboration approach among stakeholders and at all levels of government is crucial to incorporate genomics in the health system in an effective, equitable, and ethical way (Ricciardi et al. 2015).

Development of national policy statements and consistent guidelines are necessary to address all the aspects of genomics technologies, in order to ensure that individuals have access to high-quality care. However, as reported in a survey conducted in 2016 among Chief Medical Officers (CMOs) of 28 European Union (EU) Member States including Norway, a limited number of European countries implemented at that time a dedicated structural national policy for the governance of genomics (Mazzucco et al. 2017).

The application of genomics to healthcare systems has raised several concerns among the general populations, physicians, and researchers, particularly related to data privacy, safety, security, genomics literacy, and sharing of individual genomic information. Other significant issues that have to be consistently managed include the lack of data standardisation, different laboratory testing procedures, and the variable legislation that currently regulate the provision and the use of genetic tests. The potential costs of genomics to health economy are considerable and might increase unless national action is taken. Given the increasing availability of genomic technologies and their decreasing cost, many countries will be restructuring their genomic services to prepare for increasing demand. Identifying the factors that might obstacle the integration of genomics medicine into practice is challenging. Fundamental factors for a successful implementation, according to PRECeDI recommendations, are professional education in genomics, laboratory quality standards, and public awareness. While in the Chap. 3, Sect. 3.3 we discussed on the educational needs of PH professionals, in this chapter we outline the training needs of the entire category of healthcare professionals.

Genomics Educational Issues and Challenges

The exponential growth of the impact of genomics in clinical practice emphasises the clear need for a widely provision of genomics education to both current and future healthcare professionals. Although genomic education of healthcare providers has started to receive great attention in the past decades, there is still a non-conclusive evidence whether the training is adequate for the implementation of genomics in medicine (Campion et al. 2019). Current education faces some deficiencies in providing the appropriate skills, attitudes, and abilities needed. The challenges that healthcare professionals have to face every day prevent them from understanding the implications of genomics in their clinical practice. The integration of genetic/genomic relevant data into electronic health records has become increasingly important for access to genomic information in daily practice (Houwink et al. 2013, 2019; Williams et al. 2019). The biggest challenge to deal with is how to raise awareness among all healthcare professionals about the utility, benefits, and the impact of genomics application in medicine. Several articles assessing the level of knowledge among different healthcare providers indicate that general physicians do not feel sufficiently competent in genomics, in terms of interpreting a patient's genome sequencing information and making treatment decisions. Yet, not having the adequate knowledge, most of primary care providers refer patients to medical geneticists for genetic advices, overwhelming specialist genetic services. The unwillingness of general practitioners (GP) to provide their genetic counselling role has been justified by time and organisational constrains. On the other hand, even nurses think about genomics as not relevant to their role of providing optimum care to patients and their families (Mikat-Stevens et al. 2015). The lack of genomic content in the traditional curricula has negative consequences in the perception of their utility (Dotson et al. 2016), thus, in this context, one of the main policy issues to consider is the integration of genomics education into the curricula. A set of core competencies required for genomic medicine for healthcare professionals is not identified yet, whereas in genetics, a recent systematic review and a Delphi method identified three curricula for non-geneticist healthcare professionals (Tognetto et al. 2019). Nonetheless, in 2013, The National Human Genome Research Institute in the United States formulated a set of genomics competencies for physicians that could be used as a starting point for policymakers to develop a customised set of competencies. They published a framework delineating "entrustable professional activities" (EPAs) as the professional activities that together constitute the mass of critical elements that operationally define a profession specifically. Family history, genomic testing, treatment based on genomic results, somatic genomics and microbial genomic information were the identified EPAs, which comprised a set of genomic competencies and skills. Each EPA integrated the following core competencies: patient care, knowledge for practice, practice-based learning and improvement, interpersonal and communication skills, professionalism, system-based practice, interprofessional collaboration, and personal and professional development (Korf et al. 2014).

However, there is still a lack of a defined set of competencies in genomics for healthcare professionals, which might be a barrier to the development of curricula and educational opportunities as a support in achieving the necessary competencies.

Additional challenges, such as physician's literacy and the attitudes towards genomics, obstacle the adoption of genome sequencing into medical practice. The limited knowledge on genomic medicine among healthcare providers might act as a barrier to its implementation. The lack of awareness about the importance of genomics in the practice of medicine suppresses healthcare professionals' motivation to learn. Perhaps one of the main challenges to deal is the unawareness of the need to learn and the scepticism towards the potential of genomics in medicine. Notwithstanding that some professional leaders have begun to acknowledge the potential benefits of genomics, most healthcare providers have the perception that understanding genomics will not improve their practice (Berg et al. 2011).

Training Initiatives in Genomics

Recognising the need for a continuous improvement of healthcare practitioner knowledge, a number of initiatives related to genomic-based educational approaches and capacity building have been adopted at international, EU, and national level.

At international level, a major initiative called "The Global Consortium for Genomic Education" was launched, with the objective of developing a concerted programme for genomic education and training, delivered globally, with a particular focus on developing countries and less developed nations (Dhavendra Kumar 2017). Another initiative, the Training Residents in Genomics (TRIG) Working Group, has been considered as a successful strategy in building a rigorous genomic curriculum with tools for implementation aimed at postgraduate residents (Wilcox et al. 2018). Recently, the JAMA journal has launched an initiative called "JAMA Insights: Genomic and Precision Health", consisting of a series of educational articles intending to help non-geneticist clinicians to overcome knowledge barriers (Feero WG 2017).

Several educational strategies at EU level are already going in this direction. An initiative co-funded by EU called "Gen-Equip" developed a platform with the objective of training primary care practitioners in the field of genetics and genomics, through accessible online learning modules, webinars, and useful tools in daily practice. This project involved partners from six European countries: Portugal, Italy, Netherlands, the UK, Iceland, and Czech Republic, and it was available freely online in six respective languages. A study that evaluated the impact of Gen-Equip electronic resources showed that they were effective in improving genetic knowledge and skills, and helped to implement changes in clinical practice (Jackson et al. 2019). In 2018 started the initiative "The Innovative Partnership for Action Against Cancer (iPAAC) Joint Action", that brings together 44 partners from 24 European countries, aiming at developing innovative approaches to advances in cancer control, with a key focus on implementation processes among the EU-level policymakers at local, regional, and national level. Education and training on genomics for healthcare professionals is covered by a dedicated task, whose output is the implementation of an online distance course for healthcare professionals involved in cancer genomics (IPAAC Joint Action 2018).

At the national level, several initiatives were developed with the aim to achieve better genomics literacy for both healthcare professionals and the general public. However, few countries (France, The Netherlands, Croatia, Estonia, Finland, Hungary, Italy, Malta, Norway, Poland, Slovenia, Spain, and the UK) recognised the presence of pre-graduate and postgraduate university courses of genomics in healthcare (Mazzucco et al. 2017), at the survey conducted in 2016 among CMOs of 28 EU Member States and also Norway. In Italy, the pioneer in the field of policy on PH genomics, a distance learning course with an online platform (https://www.eduiss.it/), has been created, in the context of a project funded by the Italian Ministry of Health, aiming the implementation of the National Plan. The course focused on genetic/genomic testing in clinical practice, pharmacogenomics, and oncogenomics, and it was directed, in particular, to GP. It was developed in an innovative way, according to the main models of anagogical training (Problem-based Learning and Case-based Learning) (Calabrò et al. 2019).

Furthermore, in the context of a new project funded by the Italian Ministry of Health and launched in 2019, two new distance learning courses on practical genetics/omics have been created. In particular, an advanced course directed to physicians and biologists and another basic course for all other health professionals (technicians, pharmacists, nurses, etc.) have been carried out. The production of these courses responds to the achievement of the objective reported in the new National Plan to implement the capacity building of health professionals in the genetic /omics field. The two courses will be delivered through an institutional platform (https://www.eduiss.it/) and will be developed according to the main anagogic training models.

In the UK, the National Human Genome Research Institute assembled in 2013 the Inter-Society Coordinating Committee for Practitioner Education in Genomics aiming to develop and share best practices for improving the genomic literacy of providers. In 2014, they published a framework defining five entrustable professional activities in genomics, each with integrated core competencies (Korf et al. 2014).

Strategies for Integrating Genomic Science into Medical Curriculum

In the area of disruptive omics innovation, it is necessary to work on building the capacity of all health professionals. Nowadays, an appropriate and ever greater integration of omics sciences into PH and clinical practice is needed (Boccia, 2014). A continuum collaboration of experts in genomics, PH professionals, stakeholders, and healthcare practitioners is required for the establishment of common standards for education and healthcare practice. Moreover, including genomics into government priorities, developing national initiatives and appropriate guidelines and frameworks might tackle the genomic education issues (Ricciardi and Boccia 2017).

The skills and abilities required in the area of genomic medicine should be defined in order to further develop the appropriate learning methods. Considering the differences in national health systems, in terms of professional education and regulation, a standard set of competencies might guide to develop implementation strategies appropriate to the national context and further might provide an appropriate framework for genomics education across national boundaries. Despite the differences in healthcare setting of genomic application, the urge for basic notions in genomics among the practitioners is quite similar everywhere (Ricciardi et al. 2015).

The specific learning needs of healthcare practitioners should be recognised, and according to them, the curriculum should be elaborated and constructed in relation to their specific activities. It is essential to integrate in the curricula relevant topics in genomic medicine, such as PM, genome-wide association studies (GWAS), direct-to-consumer genetic testing (DTC-GT), pharmacogenomics, risk assessment, and genetic-based probability. Stakeholders should consider different approaches aiming to improve the coverage of all these topics into the current curriculum. The curriculum design and the development of the educational materials should be done on a national basis, involving genetic experts.

Incorporating genomic concepts in medical training, in particular in oncology, gynaecology, and internal medicine, may enrich the students' exposure to genomics. An added benefit of the integration of genomics with other topics within the medical curricula has to do with the application of knowledge to clinical practice. Given the novelty of genomics field, it would be relevant teaching the key concepts along with different ways of application. A continuum update of the existing curricula is essential, in order to encompass all the developments in genomics field. Education in genomics should extend across medical school or residency training programmes and it should be addressed to all healthcare professionals (Wilcox et al. 2018).

Fundamental changes are needed in the educational infrastructure in order to make education programmes accessible to every healthcare professional. A critical appraisal of the scientific literature is necessary in order to consider whether education of high quality can be provided. A high-quality education requires substantial resources, which should be flexible, updated, and readily accessible, in order to produce genomics-literate healthcare practitioners. The resources used until now into medical school curriculum or in postgraduate training programmes are insufficient for providing the education needed. This was outlined in a survey among Italian PH schools, showing that in over 60% of the schools their courses were focused on traditional genetics, and in only 47% on the impact of genomics in healthcare provision (Ianuale et al. 2014). Almost every area of the clinical medicine will be introduced with the genomic concepts, thus of utmost importance is facilitation and sharing of educational resources.

Although, over the past decade, there are several online courses on genetics/genomics topics, not all healthcare practitioners are aware of their availability and accessibility. Learners need educational approaches that are freely accessible to everyone, in various formats such as courses, online learning modules, clinical problems presentation, case series, or problem-based learning. Several educational strategies have been piloted in the United States in medical educational setting, which may serve as a model for other countries. Educational approaches are more likely to be successful if they are based on a comprehensive consideration of the aspects that are relevant for the healthcare provider practice (Pastorino et al. 2018). Innovative approaches should be developed and promoted in order to deliver the genomic information to healthcare practitioners. With the rapid advancements in genomic technologies the demand for well-instructed healthcare professionals in genomics will grow exponentially. Genomics-based knowledge will be of utmost importance for every healthcare practitioner, regardless of the field of practice; therefore, it is fundamental to promote the inevitable necessity to learn the underlying genomics. Education is undoubtedly crucial for the effective and successful implementation of genomics. Genomics education needs to evolve along with the changing scientific landscapes.

4.3   Policy Issues of Genomics in Healthcare: A Focus on Citizen Expectations and Behaviours

Michele Sassano, Giovanna Elisa Calabrò, and Stefania Boccia

The successful integration of omics information into healthcare also depends upon the actions of another central group of stakeholders, namely citizens (Budin-Ljøsne and Harris 2015). Citizens are expected to adopt a range of new behaviours and practices in relation to their healthcare. These are grouped into three overarching themes as described by Budin-Ljøsne et al. in 2015 (Budin-Ljøsne and Harris 2015), namely citizens' engagement in their own health, contribution to the research endeavour, and engagement in design and development of PM. Thus, citizens are expected to participate in the health decision-making processes, in research projects, and in debates on PM, in order to use new technologies and manage their health information, and make their contribution in collaborations with researchers, health professionals, public health authorities, and drug manufacturers through involvement in patient advocacy groups, advisory boards, and HTA bodies.

Policy Issues in Genomics: The Current Landscape for Direct-To-Consumer Genetic Tests

Over the past years, a topic has generated a huge debate among physicians, bioethicists, and government bodies: the phenomenon of DTC-GTs.

Traditionally, genetic tests were provided to patients only with the involvement of a healthcare provider, but recently various companies started to offer these tests directly to consumer. In particular, DTC-GTs can be defined as genetic tests that are both marketed and sold directly to the public, including over the internet, without the supervision of a healthcare professional (EASAC and FEAM 2012). Typically, this kind of tests is advertised and sold over the internet, and delivered to customers through mailing services, even though some of them can be found in pharmacies, supermarkets, and physical stores too. After the reception of a testing kit, the customer is required to collect and send back to the company a biological sample (e.g., hair or saliva), that will eventually be analysed. The customer is expected then to receive results either through mail or company website (Su 2013).

In recent years, together with the increasing consciousness and awareness of the public about health-related topics over time, the increasing offer and decreasing costs are favouring the growth of interest and request for DTC-GTs. Indeed, public awareness regarding DTC-GTs has been increasing over time, and as a consequence their market is growing over time and, while it was valued around US$ 824 million in 2018, its value is expected to reach a value over US$ 6 billion by 2028 ("Global Direct-to-Consumer Genetic Testing (DTC-GT) Market: Focus on Direct-to-Consumer Genetic

Testing Market by Product Type, Distribution Channel, 15 Countries Mapping, and Competitive Landscape - Analysis and Forecast, 2019-2028" 2019).

Due to the increasing offer and use and to their provision model, DTC-GTs have gathered attention by policymakers regarding potential benefits and risks.

Potential Benefits and Risks of Direct-To-Consumer Genetic Tests

Among the possible benefits, DTC-GTs could allow citizens to make informed choices about their health, and it is expected that knowing own genetic risk for a multifactorial disease could improve personal lifestyle choices, reducing healthcare costs, even though this has not yet been clearly proved (Su 2013). Actually, evidence suggests that DTC-GT results may have only a modest impact on lifestyle changes (Covolo et al. 2015).

However, DTC-GTs raise many concerns too, such as information privacy and security. In particular, undergoing a genetic test let companies gather information not only about the individual, but also about other family members, even not yet born ones, which could affect their lives at many levels, such as insurance premium, employment, and relationships. In particular, most companies declare strict policies regarding data privacy, even though often an easily accessible privacy policy document is not available on their websites, and when it is, in many cases it does not explain exactly possible uses of individual's data and details about methods used to ensure data privacy and security. In addition, most companies do not inform consumers directly and individually in case of privacy policy changes (Hazel and Slobogin 2018). The importance of privacy of sensitive genetic data is further underlined by the Statement of the European Society of Human Genetics (ESHG) on DTC-GTs (Borry 2010).

As described in the Statement of the ESHG, a relevant issue is related to the predictive ability of genetic variants evaluated by DTC-GTs, which could compromise their clinical validity and utility (Borry 2010). In particular, premature commercialisation of these tests should be avoided, while strong evidence of their validity should be provided before placing them on the market (Borry 2010) in order to assure validity of tests and their results and to avoid false positives, which have been estimated being around 40%, and misinterpretation of the risk associated with analysed variants (Tandy-Connor et al. 2018). In the end, transparency regarding test validity by DTC-GT selling companies should be pursued, though stricter regulation of the methods used to prove validity and of advertising and promotional materials. To this end, **PRECeDI recommendation n. 1** focuses on the need of strong evidence of efficacy and/or effectiveness of a new technology, such as biomarkers, genetic testing, and personalised interventions aimed at the prevention of chronic diseases to allow its implementation in healthcare (Boccia et al. 2019).

Another drawback of DTC-GTs regards the ability of health professionals to understand and interpret the results (Su 2013). In particular, health professionals lack competencies needed to support patients understanding genetic risk information, showing a need for a better education in this field (Covolo et al. 2015). This is particularly relevant for primary care physicians, being most often the first line in patient information and care in European Countries. A better understanding of general principles regarding DTC-GTs should be pursued, in order to allow health professionals assist their patients in an appropriate way, avoiding waste of healthcare resources. Furthermore, knowing about personal genetic risk for a disease could make consumers experience psychological distress, and this could be further accentuated by the lack of a clear explanation of testing results by companies and/or health professionals. Even though some experts assert DTC-GT results carry little or no psychological consequence, a clear and definitive explanation has not yet been provided (Su 2013). To allow citizens receive correct information and avoid the abuse of further diagnostic investigation or psychological distress, it is essential to provide pre-test and post-test genetic counselling, since company websites cannot replace face-to-face counselling with an expert professional (Borry 2010). To this end, **PRECeDI recommendation n. 5** underlines the importance of the identification of an efficient provision model for genetic tests, in which the involvement of different healthcare professionals and new professional roles is essential. Professional education/training in genomics medicine, laboratory quality standards, and public awareness are essential factors for the successful implementation of genomic applications in practice (Boccia et al. 2019). Furthermore, ESHG underlines the importance of impartiality of health advices by consultants, raising concerns about them in the cases in which the professional is employed or anyway linked to the companies selling the tests (Borry 2010). Eventually, if consumers are not offered adequate pre-test and post-test counselling and psychological support when needed, and tests are not clinically valid and useful, tests cannot be used to interpret personal genetic risk and make recommendations to promote individuals' health (Borry 2010). The importance of the involvement of a health professional, in particular in the interpretation of DTC-GT results, is further underlined by statements of several American and European medical and scientific associations (Skirton et al. 2012; Rafiq et al. 2015). To this end, citizen education is essential to avoid false hopes and fears related to DTC-GT results and to foster the search for medical supervision in this field (Borry 2010). For appropriate and effective educational programmes and strategies, it is important to understand the citizen's perceptions. With this aim, a recent systematic review identified the European citizens' perspectives towards DTC-GTs and reported an overall low level of knowledge on these tests and a high interest in purchasing them (Hoxhaj et al. 2020a).

A further key issue is related to consent, particularly for minors and for research purposes. First, most DTC-GT selling companies only require to sign an informed consent form, which actually shouldn't replace a correct informed consent process needed for consumers to really understand information. Moreover, consumers are expected to collect samples at home and mail them to the company for analysis, but this process does not ensure the provider of the biological sample is really the claimed one (Borry 2010). Furthermore, often companies use data for biomedical research, but not always there is a clear consent request and explanation for this purpose. Eventually, as for minors, ESHG states that, since minors cannot give a correct informed consent, DTC-GTs should not be undertaken in this particular population, while testing should be allowed only if ordered by a qualified health professional (Borry 2010).

Eventually, a systematic review of position statements, policies, guidelines, and recommendations, produced in Europe (Rafiq et al. 2015), showed that professional societies and associations are currently more suggestive of potential disadvantages of DTC-GTs. Considering the difficulty in creating international standards that regulate the online market, the scientific community underlined the need to promote an agreement on a code of practice based on specific recommendations that include appropriate genetic literacy of both populations and health professionals, implementation research on the genetic tests to integrate PHG into healthcare systems, as well as the guarantee of appropriate information to consumers (Skirton et al. 2012; Rafiq et al. 2015).

Oversight and Regulation

Oversight of DTC-GTs in the United States is the responsibility of the Food and Drug Administration (FDA), which considers DTC-GTs with medical indications as medical devices, hence requiring regulatory clearance. In November 2013, the United States FDA ordered 23andMe, a provider of DTC genomic services, to stop marketing health-related genetic tests due to the risk that false results could cause consumers to undergo unnecessary health procedures (Yim and Chung 2014). However, later FDA changed its position regarding this kind of tests. In particular, in 2015 FDA authorised for marketing 23andMe's Bloom syndrome (a rare autosomal recessive disorder) carrier test, while in 2017 FDA allowed the marketing of the first DTC-GTs providing genetic risk information for certain non-hereditary conditions, such as Parkinson's, Alzheimer's, and celiac diseases. Eventually, in 2018 a test for three BRCA mutations was authorised for marketing by FDA.

As stated above, many concerns are related to privacy and discriminatory consequences of DTC-GT results. In this context, the United States released the Genetic Information Nondiscrimination Act (GINA) in 2008 in order to protect consumers from discrimination in the fields of health insurance and employment derived from genetic testing results. However, this law is far to be perfect, since it does not cover disability insurance, life insurance, and long-term care insurance (Borry 2010).

As described above, another relevant issue is linked to the validity of the scientific background behind DTC-GTs. First of all, a test can be considered valid if it achieves analytical validity and clinical validity, even if clinical utility is a relevant parameter too. Regarding analytical validity, most DTC-GT providing laboratories based in the United States are aligned to the Clinical Laboratory Improvement Amendments (CLIA), which are designed to control the quality of laboratory practices. On the other hand, it is extremely difficult to establish clinical validity, which refers to the strength of the association between the genetic variant being analysed and the presence or risk of a specific disease, and clinical utility, which refers to the ability of the test to provide information that will be useful in the care process (Genetics Home Reference 2016).

Since most DTC-GT providing laboratories are based in the United States, federal laws in this Country are much more focused on this kind of tests than European ones. In particular, specific laws addressing DTC-GTs lack in European Member States, while laws with other targets can affect this field, bringing to overlapping or missing of regulation in some cases (Kalokairinou et al. 2018). For instance, the Additional Protocol to the Convention on Human Rights and Biomedicine prescribes the need of medical supervision in case of genetic tests for health purposes, while not all European States adhere to this prescription, with some Countries requiring mandatory medical supervision or genetic counselling for different testing types. Furthermore, Countries have different specific requirements regarding informed consent for genetic testing (Kalokairinou et al. 2018). Eventually, EU General Data Protection Regulation (GDPR) legislation will affect the way companies handle consumer's data (Friend et al. 2018), even though it does not address specifically DTC-GTs.

However, to ensure appropriateness of regulation and avoid premature introduction of genetic tests into the healthcare system in European Countries, it is essential to adapt, when needed, and implement In Vitro Diagnostic Devices Directive, Additional Protocol to the Convention on Human Rights and Biomedicine, and OECD Council Recommendation on Quality Assurance in Molecular Genetic Testing, even though they're not specific for DTC-GTs (Borry 2010).

In conclusion, DTC-GTs appear as an attractive field both for companies, due to economic growth possibilities, and for citizens, who can easily access personal genetic information regarding health-related or not related conditions. However, it is important to assess limitations and ethical, legal, and social implications (ELSI) surrounding tests, and to make consumers aware of them, in order to let them make informed decisions. Moreover, even though it has become extremely difficult due to the rapid growth of the market, further regulation of DTC-GTs is desirable in order to protect users' rights. As for Europe, a recent literature review that summarised the legislations related to DTC-GTs among the EU Member States reported that specific legislative instruments on DTC-GTs have not been implemented yet at EU level or at a national level (Hoxhaj et al. 2020b). Therefore, the lack of a unified regulatory framework indicates the need for common laws and rules that should apply in EU level. In the end, ethico-legal and policy issues are targeted by **PRECeDI recommendation n. 3**, which underlines the importance of obligations and responsibilities of researchers, research participants, and general public too (Boccia et al. 2019).

Conclusions

In order to accomplish all these requirements and expectations, citizens need to be educated and motivated in changing their usual approach to healthcare.

The appropriate and conscious utilisation of new technologies and services by citizens requires correct information of users, regarding not only DTC-GTs, but also omics and genomics sciences and their possibility and limitations. For instance, in the Italian context, this is underlined by authorities with the first "Guidelines on genomics in public health" of 2013 (Simone et al. 2013) and the "National Plan for innovation of the Health System based on omics sciences" of 2017 (Boccia et al. 2017). In both cases, the importance of raising health professionals', policymakers', leaders', and citizens' awareness is highlighted and considered a need for the implementation of omics science in the healthcare system.

The data we reviewed reveal that citizens are expected to adopt a whole range of behaviours and practices considered to be crucial for the realisation of PM. In this context, the role of citizens compared with the way they traditionally have been involved in their healthcare has radically changed. This change is related to recent developments which encourage a move away from the paternalistic model, under which citizens are primarily passive recipients of healthcare, to a participatory model of healthcare under which citizens are responsible drivers of their health, contributors to the healthcare system, and partners sharing decisions with healthcare providers (Ricciardi and Boccia 2017). This new role of citizens offers exciting opportunities but requires levels of health and technology literacy as well as adequate socioeconomic resources. Education in PM is a priority.

Citizens will be more receptive in adopting new behaviours and practices and contribute to the realisation of PM only if educational, socioeconomic, and cultural hurdles are correctly addressed.

References

Adjekum, A., Ienca, M., & Vayena, E. (2017). What is trust? Ethics and risk governance in precision medicine and predictive analytics. *OMICS A Journal of Integrative Biology, 21*(12), 704–710.

Appelbaum, P.S., Roth, L.H., Lidz, C.W., et al. (2018) *False hopes and best data: Consent to research and the therapeutic misconception*. Research Ethics. Routledge: 167–171.

Berg, J. S., Khoury, M. J., & Evans, J. P. (2011). Deploying whole genome sequencing in clinical practice and public health: Meeting the challenge one bin at a time. *Genetics in Medicine, 13*(6), 499–504.

Blasimme, A., & Vayena, E. (2016). Becoming partners, retaining autonomy: Ethical considerations on the development of precision medicine. *BMC Medical Ethics, 17*(1), 67.

Boccia, S. (2012). Personalized health care: The hope beyond the hype. *Italian Journal of Public Health, 9*(4).

Boccia, S. (2014). Why is personalized medicine relevant to public health? *European Journal of Public Health, 24*(3), 349–350.

Boccia, S., Federici, A., Siliquini, R., et al. (2017). Implementation of genomic policies in Italy: The new National Plan for innovation of the Health System based on omics sciences. *Epidemiology, Biostatistics and Public Health, 14*(4).

Boccia, S., Pastorino, R., Ricciardi, W., et al. (2019). How to integrate personalized medicine into prevention? Recommendations from the personalized prevention of chronic diseases (PRECeDI) consortium. *Public Health Genomics, 22*(5-6), 208–214.

Bombard, Y., Brothers, K. B., Fitzgerald-Butt, S., et al. (2019). The responsibility to recontact research participants after reinterpretation of genetic and genomic research results. *American Journal of Human Genetics, 104*(4), 578–595.

Borry, P. (2010). Statement of the ESHG on direct-to-consumer genetic testing for health-related purposes. *European Journal of Human Genetics, 18*(12), 1271–1273.

Budin-Ljøsne, I., & Harris, J. R. (2015). Ask not what personalized medicine can do for you - Ask what you can do for personalized medicine. *Public Health Genomics, 18*(3), 131–138.

Calabrò, G. E., Tognetto, A., Mazzaccara, A., et al. (2019). Omic sciences and capacity building of health professionals: A distance learning training course for Italian physicians, 2017-2018. *Igiene e sanita pubblica, 75*(2), 105–124.

Campion, M., Goldgar, C., Hopkin, R. J., et al. (2019). Genomic education for the next generation of health-care providers. *Genetics in Medicine, 21*(11), 2422–2430.

Collins, F. S., & Varmus, H. (2015). A new initiative on precision medicine. *New England Journal of Medicine, 372*(9), 793–795.

Covolo, L., Rubinelli, S., Ceretti, E., et al. (2015). Internet-based direct-to-consumer genetic testing: A systematic review. *Journal of Medical Internet Research, 17*(12), e279.

David, K. L., Best, R. G., Brenman, L. M., et al. (2019). Patient re-contact after revision of genomic test results: Points to consider—a statement of the American College of Medical Genetics and Genomics (ACMG). *Genetics in Medicine, 21*(4), 769–771.

Kumar, D. (2017) The Global Consortium for Genomic Education. Retrieved from https://www.genomicmedicine.org/global-consortium-genomic-education/. Accessed 25 Sep 2019.

Dotson, W. D., Bowen, M. S., Kolor, K., et al. (2016). Clinical utility of genetic and genomic services: Context matters. *Genetics in Medicine, 18*(7), 672–674.

Easter, M. M., Henderson, G. E., Davis, A. M., et al. (2006). The many meanings of care in clinical research. *Sociology of Health & Illness, 28*(6), 695–712.

EASAC and FEAM (2012) Direct-to-consumer genetic testing for health-related purposes in the European Union. Report. *Available from https://easac.eu/fileadmin/Reports/Easac_12_DTC_Web.pdf.*

Evans, J. P., Meslin, E. M., Marteau, T. M., et al. (2011). Deflating the genomic bubble. *Science, 331*(6019), 861–862.

Feero, W. G. (2017). Introducing "Genomics and Precision Health". *JAMA, 317*(18), 1842–1843.

Fiore, R. N., & Goodman, K. W. (2016). Precision medicine ethics: Selected issues and developments in next-generation sequencing, clinical oncology, and ethics. *Current Opinion in Oncology, 28*(1), 83–87.

Friend L, Rivlin A, O'Neill J et al. (2018) Direct-to-consumer genetic testing: Opportunities and risks in a rapidly evolving market. *KPMG International* 12. Retrieved from https://assets.kpmg/content/dam/kpmg/xx/pdf/2018/08/direct-to-consumer-genetic-testing.pdf. Accessed 13 Dec 2019.

Genetics Home Reference. (2016). How can consumers be sure a genetic test is valid and useful? Retrieved from https://ghr.nlm.nih.gov/primer/testing/validtest Accessed 13 Dec 2019.

Global Direct-to-Consumer Genetic Testing (DTC-GT) Market: Focus on Direct-to-Consumer Genetic Testing Market by Product Type, Distribution Channel, 15 Countries Mapping, and Competitive Landscape - Analysis and Forecast, 2019–2028. (2019). Retrieved from https://www.researchandmarkets.com/reports/4771288/global-direct-to-consumer-genetic-testing-dtc?utm_source=CI&utm_medium=PressRelease&utm_code=xbhgsj&utm_campaign=1249731+-+Direct-to-Consumer+Genetic+Testing+(DTC-

GT)+Market%2C+2028+-+Partnerships+with+Different+Distribution+Channels+and+Health+Insurance&utm_exec=chdo54prd.

Green, R. C., Berg, J. S., Grody, W. W., et al. (2013). ACMG recommendations for reporting of incidental findings in clinical exome and genome sequencing. *Genetics in Medicine, 15*(7), 565–574.

Hammer, M. J. (2016). Precision medicine and the changing landscape of research ethics. *Oncology Nursing Forum, 43*(2), 149–150.

Hazel, J., & Slobogin, C. (2018). Who knows what, and when?: A survey of the privacy policies proffered by U.S. Direct-to-Consumer Genetic Testing Companies. *Cornell Journal of Law and Public Policy, 28*, 35.

Henry, L. M. (2013). Introduction: Revising the common rule: Prospects and challenges. *Journal of Law, Medicine and Ethics, 41*(2), 386–389.

Houwink, E. J., Sollie, A. W., Numans, M. E., et al. (2013). Proposed roadmap to stepwise integration of genetics in family medicine and clinical research. *Clinical and Translational Medicine, 2*(1), 5.

Houwink, E. J., Hortensius, O. R., van Boven, K., et al. (2019). Genetics in primary care: Validating a tool to pre-symptomatically assess common disease risk using an Australian questionnaire on family history. *Clinical and Translational Medicine, 8*(1), 17.

Hoxhaj, I., Stojanovic, J., & Boccia, S. (2020a). European citizens' perspectives on direct-to-consumer genetic testing: An updated systematic review. *European Journal of Public Health, 63*(4), 103841. https://doi.org/10.1093/eurpub/ckz246.

Hoxhaj, I., Stojanovic, J., Sassano, M., et al. (2020b). A review of the legislation of direct-to-consumer genetic testing in EU member states. *European Journal of Medical Genetics*. https://doi.org/10.1016/j.ejmg.2020.103841.

Ianuale, C., Leoncini, E., Mazzucco, W., et al. (2014). Public Health Genomics education in post-graduate schools of hygiene and preventive medicine: A cross-sectional survey. *BMC Medical Education, 14*(1), 213.

IPAAC Joint Action. (2018). Retrieved from https://www.ipaac.eu/.

Jackson, L., O'Connor, A., Paneque, M., et al. (2019). The Gen-Equip Project: Evaluation and impact of genetics e-learning resources for primary care in six European languages. *Genetics in Medicine, 21*(3), 718–726.

Juengst, E., McGowan, M. L., Fishman, J. R., et al. (2016). From "personalized" to "precision" medicine: The ethical and social implications of rhetorical reform in genomic medicine. *Hastings Center Report, 46*(5), 21–33.

Kalokairinou, L., Howard, H. C., Slokenberga, S., et al. (2018). Legislation of direct-to-consumer genetic testing in Europe: A fragmented regulatory landscape. *Journal of Community Genetics, 9*(2), 117–132.

Knoppers, B. M., Thorogood, A., & Zawati, M. H. (2019). Letter: Relearning the 3 R's? Reinterpretation, recontact, and return of genetic variants. *Genetics in Medicine, 11*, 1.

Korf, B. R., Berry, A. B., Limson, M., et al. (2014). Framework for development of physician competencies in genomic medicine: Report of the Competencies Working Group of the Inter-Society Coordinating Committee for Physician Education in Genomics. *Genetics in Medicine, 16*(11), 804–809.

Kraft, S. A., Cho, M. K., Gillespie, K., et al. (2018). Beyond consent: Building trusting relationships with diverse populations in precision medicine research. *American Journal of Bioethics, 18*(4), 3–20.

Mazzucco, W., Pastorino, R., Lagerberg, T., et al. (2017). Current state of genomic policies in healthcare among EU member states: Results of a survey of chief medical officers. *European Journal of Public Health, 27*(5), 931–937.

Mikat-Stevens, N. A., Larson, I. A., & Tarini, B. A. (2015). Primary-care providers' perceived barriers to integration of genetics services: A systematic review of the literature. *Genetics in Medicine, 17*(3), 169–176.

Pastorino, R., Calabrò, G. E., Lagerberg, T., et al. (2018). Effectiveness of educational intervention types to improve genomic competency in non-geneticist medical doctors: A systematic review of the literature. *Epidemiology Biostatistics and Public Health, 15*(1), e12657-1–e12657-8.

Rafiq, M., Ianuale, C., Ricciardi, W., et al. (2015). Direct-to-Consumer genetic testing: A systematic review of european guidelines, recommendations, and position statements. *Genetic Testing and Molecular Biomarkers, 19*(10), 535–547.

Ricciardi, W., Villari, P., McKee, M., et al. (2015). Public Health Genomics: Moving towards the implementation of dedicated policies in Europe. *European Journal of Public Health, 25*(suppl_3), ckv169-023.

Ricciardi, W., & Boccia, S. (2017). New challenges of public health: Bringing the future of personalised healthcare into focus. *European Journal of Public Health, 27*(suppl_4), 36–39.

Sankar, P. L., & Parker, L. S. (2017). The Precision Medicine Initiative's All of Us Research Program: An agenda for research on its ethical, legal, and social issues. *Genetics in Medicine, 19*(7), 743–750.

Schleidgen, S., Klingler, C., Bertram, T., et al. (2013). What is personalized medicine: Sharpening a vague term based on a systematic literature review. *BMC Medical Ethics, 14*(1), 55.

Schuol, S., Schickhardt, C., Wiemann, S., et al. (2015). So rare we need to hunt for them: Reframing the ethical debate on incidental findings. *Genome Medicine, 7*(1), 83.

Simone, B., Mazzucco, W., Gualano, M. R., et al. (2013). The policy of public health genomics in Italy. *Health Policy, 110*(2–3), 214–219.

Skirton, H., Goldsmith, L., Jackson, L., et al. (2012). Direct to consumer genetic testing: A systematic review of position statements, policies and recommendations. *Clinical Genetics, 82*(3), 210–218.

Su, P. (2013). Direct-to-consumer genetic testing: A comprehensive view. *The Yale Journal of Biology and Medicine, 86*(3), 359–365.

Tandy-Connor, S., Guiltinan, J., Krempely, K., et al. (2018). False-positive results released by direct-to-consumer genetic tests highlight the importance of clinical confirmation testing for appropriate patient care. *Genetics in Medicine, 20*(12), 1515–1521.

Thorogood, A., Dalpé, G., & Knoppers, B. M. (2019). Return of individual genomic research results: Are laws and policies keeping step? *European Journal of Human Genetics, 27*(4), 535–546.

Tognetto, A., Michelazzo, M. B., Ricciardi, W., et al. (2019). Core competencies in genetics for healthcare professionals: Results from a literature review and a Delphi method. *BMC Medical Education, 19*(1), 19.

Wilcox, R. L., Adem, P. V., Afshinnekoo, E., et al. (2018). The Undergraduate Training in Genomics (UTRIG) Initiative: early & active training for physicians in the genomic medicine era. *Personalized Medicine, 15*(3), 199–208.

Whitcomb, D. C. (2012). What is personalized medicine and what should it replace? *Nature Reviews Gastroenterology and Hepatology, 9*(7), 418.

Williams, M. S., Taylor, C. O., Walton, N. A., et al. (2019). Genomic information for clinicians in the electronic health record: Lessons learned from the clinical genome resource project and the electronic medical records and genomics network. *Frontiers in Genetics, 10*, 1059.

Yim, S. H., & Chung, Y. J. (2014). Reflections on the US FDA's warning on direct-to-consumer genetic testing. *Genomics & Informatics, 12*(4), 151.

Yuan, A. (2019). Blurred lines: The collapse of the research/clinical care divide and the need for context-based research categories in the revised common rule. *Food and Drug Law Journal, 74*, 46.

Chapter 5
Roles and Responsibilities of Stakeholders in Informing Healthy Individuals on Their Genome: A Sociotechnical Analysis

Martina C. Cornel, Tessel Rigter, and Carla G. van El

5.1 Responsible Translation of Innovations in the Context of Screening of Healthy Adults

Martina C. Cornel, Tessel Rigter, and Carla G. van El

Background

Although most genetic disorders cannot be prevented, in the sense that the genetic variant that is causative of the disease remains present, secondary or tertiary preventative measures can often impact prognosis, i.e. symptoms can often be avoided. Identifying individuals at risk of genetic disorders among a healthy population, often referred to as screening, could therefore improve health outcomes.

Examples of successful screening approaches for genetic conditions in public health include newborn screening for phenylketonuria, where a diet (mainly phenylalanine free) could prevent children from developing symptoms such as profound cognitive impairment (Brosco and Paul 2013). Another approach where substantial health gain could be achieved is so-called "cascade screening," where the structure of the pedigree of an index case with, e.g., cancer or CVDs is used to identify family members at risk (Fig. 5.1). Offering them genetic testing could allow for personalised prevention and thereby impact health outcomes drastically (Ned and Sijbrands 2011; Henneman et al. 2013).

M. C. Cornel · T. Rigter · C. G. van El (✉)
Amsterdam University Medical Centers, location VUMC, Vrije Universiteit Amsterdam, Department of Clinical Genetics, Section Community Genetics, and Amsterdam Public Health Research Institute, Amsterdam, The Netherlands
e-mail: mc.cornel@amsterdamumc.nl; t.rigter@amsterdamumc.nl; cg.vanel@amsterdamumc.nl

© The Editor(s) and Author(s), under exclusive license to Springer Nature Switzerland AG 2021
S. Boccia et al. (eds.), *Personalised Health Care*, SpringerBriefs in Public Health, https://doi.org/10.1007/978-3-030-52399-2_5

New opportunities for effective application of genetic testing are arising constantly. Over the last decade, developments in primarily genomic technology, but also, e.g. shifts towards more patient-centred care, have induced changes in (public) healthcare practice. Policymakers and (public) healthcare providers are being challenged to adapt to these innovations, in order to make (patients in) the general public profit from their potential. This "translational process," comprises different steps in the evolution from a new invention to a useful application (Khoury et al. 2007). Structured organisation of the steps in this process could ensure responsible innovation by achieving efficient, effective, and sustainable implementation.

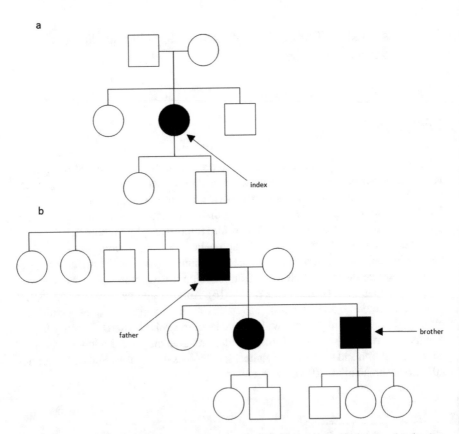

Fig. 5.1 Cascade screening. After an index patient has been identified with the disorder due to a pathogenic DNA-variant (**a**), her first-degree relatives (parents, brothers, sisters, children) are informed about DNA testing. If the father and brother also test positive (**b**), a next circle of first-degree relatives is informed; for instance, the brothers and sisters of the father and the children of the brother

Prioritisation of Innovations to Use Resources for Achieving High Value

New tools and technologies that are applied without the evidence demonstrating that they are useful run the risk of posing harm to individuals, families, and the broader health system. This includes the risk and consequences of over-, under-, or misdiagnosis, or psychological harm. A first step in responsible implementation of innovations is therefore to prioritise interventions that result in the most health gain with the least negative impact with the finite resources available.

Guidance for evaluation of predictive genomic applications has been developed for, and applied by, policymakers (Pitini et al. 2018). It should be realised that a test alone does not have inherent utility: even if it measures the intended outcomes reliably, it is the adoption of therapeutic or preventive interventions based on these outcomes that determine the value. Details on methods and challenges for the evaluation process have been elaborated in Chap. 2 "Evaluation of Predictive Genomic Applications" of this book.

In the case of screening healthy individuals, the highest priority will most likely be with populations that have a high a priori risk and where effective interventions are available (Severin et al. 2015). Some argue for instance that population screening for inherited breast and ovarian cancer may be justifiable in populations with a high carrier frequency of disease-causing variants in the genes *BRCA1* and *BRCA2*, such as the Ashkenazi Jewish population (Gabai-Kapara et al. 2014) (Box 5.1).

Box 5.1: Screening in populations with a high carrier frequency: an example
In the Ashkenazi Jewish population female carriers have a cumulative risk of developing either breast or ovarian cancer by age 80 of 83% for *BRCA1* carriers and 76% for *BRCA2* carriers. These high-risk women can opt for prophylactic surgeries and/or surveillance by MRI/mammography. Salpingo-oophorectomy around the age of 35–40 years would prevent most ovarian cancers. Thus, after a presymptomatic DNA diagnosis, cancers might be prevented and/or detected at an earlier stage.

While population screening in this specific example might be justifiable, clinical utility would be even higher in family members of known mutation carriers, since first-degree relatives have a 50% a priori risk of carrying the same mutation, while the prevalence in the Ashkenazi Jewish population is 2.5% (Roa et al. 1996).

Internationally the Wilson and Jungner criteria are used as the basis of selection for disorders to include in population screening (Andermann et al. 2008). These criteria include that a treatment is available and advantages outweigh risks. It is clear that female carriers of *BRCA1* and *BRCA2* mutations profit from early detection. In first-degree relatives the risk of having inherited a deleterious mutation is 50%, and therefore these would profit most from screening and interventions.

After evaluation and prioritisation innovation will not occur automatically. Change processes are required for full implementation of the selected interventions. It is known, however, that within the field of genomic medicine research is rarely focusing on the processes beyond the first step of developing innovations (Khoury et al. 2007). Better understanding of implementation science is therefore essential (Roberts et al. 2017).

A Theoretical Framework for a Sociotechnical Analysis

To unravel the changes that are required for responsible innovation in healthcare, or in this case more specifically: implementation of new (or adapted) genetic (public) health services, we have adopted elements of models and concepts used in the field of Health System Innovation and Transition. Central in this perspective is the notion that, in medicine, a group of individuals or actors (professionals and patients) are used to working in a certain structure, culture, and practice (see Fig. 5.2). Together, the specific ways of organising, thinking, and doing, define "the constellation: a set of interrelated practices and relevant structuring elements that together are both defining and fulfilling a function in a larger societal system in a specific way" (van Raak 2010). Technological innovation or other dynamics within or outside the constellation might induce changes by requiring actors to adapt.

In order to achieve full transitions, it is essential to realise that actors need to change in all aspects of the constellations (culture, structure, and practice) and often roles and responsibilities need to be redistributed. Therefore, explicit study of the barriers and facilitators for change and the different expectations of roles and responsibilities in each phase of a transition has proven insightful and may provide useful tools for stakeholders involved in such transitions.

Attuning All Actors in a Transition Process

Furthermore, all actors need to be attuned in a translational process to achieve full integration of innovation. The study of prerequisites for change, therefore, can only provide a complete overview if all stakeholders (and their liaisons) are taken into account. Dynamics can occur by technological developments by scientists, has to be implemented in (public) healthcare by healthcare professionals, will depend on demand of patients

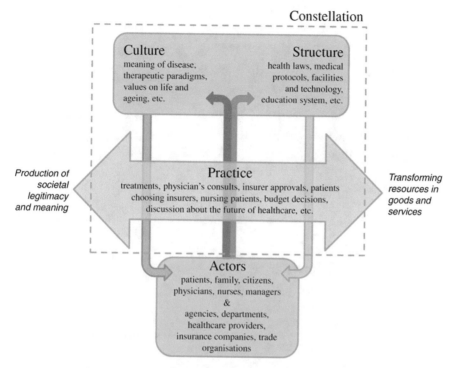

Fig. 5.2 The constellation perspective. Operationalisation of the constellation concept into structure, culture, practice, and associated actors (Rigter et al. 2014; adapted from van Raak 2010, with permission from VU University Press)

and/or citizens and will have to accommodate to ethical, legal, and social standards developed by advisory or governmental agencies (see Fig. 5.3).

Understanding Change Processes: The Transition Management Perspective

General lessons from transition management can be adapted and could aid in understanding and managing the steps in the change processes in genomic innovation. Figure 5.4 elaborates on what is needed for increased structural uptake of new interventions.

Often changes start at small scale, at local or specific sub-group level. Within these "niches" actors are exploring the potential of the innovation by learning from the first applications. Multiple dimensions of the innovation (e.g. efficacy, effectiveness, safety, acceptability, efficiency, feasibility of implementation and costs (Khoury et al. 2008)) can be addressed in different niches, each with their own perspective. During this process, actors are often reaching out to attune to like-minded actors with similar initiatives, also referred to as broadening implementation.

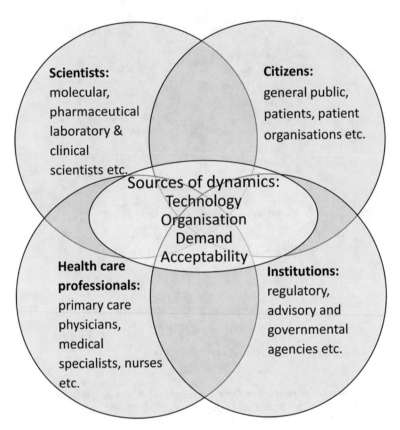

Fig. 5.3 Network of actors in transitions. Network of actors that need to be attuned in transition processes that could be initiated by dynamics in technology, organisation, demand, and/or acceptability in healthcare systems (Rigter et al. 2014; adapted from Achterbergh et al. 2007, with permission from Elsevier Ireland Ltd., https://doi.org/10.1016/j.healthpol.2007.02.007)

When this leads to collaboration in, often also more interdisciplinary, initiatives (or "patches of regimes") "deepening" of the implementation can take place by more structured learning from the application in different settings. By doing so, acceleration of implementation can take place.

When general lessons are being learned from the patches of regimes, stabilisation can occur, reaching full integration of the innovation in the constellation. This "scaling up" of implementation requires true changes in thinking, organising, and doing of the actors, and is often accelerated by a specific window of opportunity (Van den Bosch, 2010). Windows of opportunity often arise from factors outside the constellation under study, e.g. through political or social developments which lead to provision of financial incentives and/or (political) attention. Besides enough support from the key stakeholders to achieve consensus on best practice and a collective sense of urgency, the division of roles and responsibilities requires specific attention in this last phase of implementation. Although in some cases this will naturally take

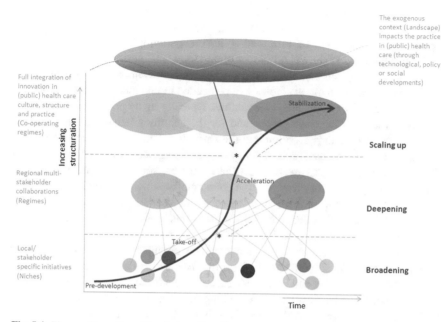

Fig. 5.4 Phases of transitions. True integration of innovation in (public) healthcare requires different phases in transition. (Rigter et al. 2020; adapted from Geels and Schot 2007, with permission from Elsevier B.V., https://doi.org/10.1016/j.respol.2007.01.003, Rotmans et al. 2001, with permission from ICIS, and Geels 2002, with permission from Elsevier Science B.V., https://doi.org/10.1016/S0048-7333(02)00062-8)

place, often this requires a "change agent" who is sensitive to and valued and trusted by all actors or key stakeholders (Essink, 2012). A change agent could establish coalitions and transparent organisational structures and help identify the required new competencies (knowledge, attitude, and skills) from the different actors. This is crucial because in order to truly scale up new services, also legal and financing frameworks should be organised.

5.2   Examples of Informing Family Members of Cardiovascular Diseases and Hereditary Cancers

Martina C. Cornel, Tessel Rigter, and Carla G. van El

Finding Index Cases and Informing Family Members Require Different Ways of Doing, Thinking and Organising

PH initiatives often target the population at large, but in the examples that we will discuss in this section, we argue that first-degree relatives of people with a monogenic subtype of certain CVDs and hereditary cancers are a population to target first

(Van El and Cornel, 2011). They have a 50% a priori risk of carrying the high-risk gene variant, and with little resources a high impact in terms of life years saved can be achieved if they will be informed about preventive interventions and will be under surveillance. While PH initiatives often target "all" persons (all newborns for neonatal screening, all persons aged 55–75 years for colon cancer screening), the structured approach in our examples will be needed to find index cases or inform all first-degree relatives. Screening programmes typically invite people to a healthcare service before they have symptoms, and for the healthy relatives that applies too. The shift from care to prevention and the shift from "all" to first-degree relatives require a different mindset (culture), way of organising (structure) and implementation in day-to-day work (practice). In the examples, some small-scale initiatives need scaling up (e.g. from one city to a country), for which governance is needed. Relevant questions will surface, for instance, about whether there is a change agent? Which stakeholder takes the lead? Which other stakeholders are needed, and what should their roles and responsibilities be?

We will look at several strategies and recent developments in various countries to find index cases and subsequently inform family members of their risk for monogenic subtypes of CVDs and hereditary cancer. From these examples we will identify facilitators and barriers for implementing or optimising cascade screening and highlight roles and responsibilities of stakeholders in changing culture, structure, and practice.

In the first case example we will address ways to optimise case finding in specialised care where genetic and non-genetic healthcare professionals need to collaborate. We will discuss the implementation of new guidelines to screen tumour tissue for LS, a form of hereditary colorectal cancer. After screening, DNA testing can confirm the diagnosis in an index case. Subsequently, healthy relatives can be informed about healthcare services. In the second case, ways to improve the identification of index patients in primary care will be addressed. We will discuss recent initiatives in the United Kingdom for systematic case finding in a hereditary form of high cholesterol, Familial Hypercholesterolaemia (FH).

In the third case, the tracing, informing, and genetic testing of family members will be addressed. In the Netherlands, a national screening programme was implemented allowing for a structured approach towards informing family members. We will discuss the challenges healthcare professionals faced after the ending of the official screening programme to still reach family members. In this way we can grasp what tasks need to be performed to ensure genetic testing can fulfil its potential for prevention.

In the Discussion (Sect 5.3) we will outline the lessons to be drawn from these cases with special attention for the role of governance and PH or professional policy to address the complete cycle from identification of index cases to cascade screening of family members, relying on close collaboration between primary and secondary care and genetic and non-genetic healthcare professionals.

Lynch Syndrome

A key element of prevention is finding cases before health damage occurs. Unfortunately, in genetic disorders usually a first case in which serious health problems occur triggers further investigations. Especially in a situation where genetic testing is expensive, a combination of clinical symptoms and family medical history may help to select cases for which genetic testing seems indicated, because there is a high likelihood to find a pathogenic gene variant. In the case of LS, a form of hereditary colorectal cancer (CRC), in the past decade new ways to facilitate and increase the number of identified index cases have gained ground. LS is responsible for 2–4% of CRC, affecting especially young patients, and is caused by dominantly inherited gene variants involved in mismatch repair mechanisms (*MLH1, MSH2, MSH6, PMS2,* and *EpCAM*). Patients have a high recurrence risk of tumours, which, if detected in time, can be removed. After the first tumour occurred, biennial colonoscopy, for instance, can help to identify a second colon tumour at an early stage, or even allow for identification and excision of adenomas, often the prodromal condition, before cells become malignant. LS also increases the risk of other types of cancer, such as endometrial, stomach, prostate, breast, and ovarian cancer. In women, surveillance by a gynaecologist and/or prophylactic hysterectomy can reduce morbidity and mortality.

Currently LS is still underdiagnosed, because until recently strategies were cumbersome. There was a lack of coordination, and referrals to geneticists depended on oncologist or surgeon assessment of pathology. Often the Amsterdam and (revised) Bethesda criteria were used that combined clinical criteria such as tumour location and age of onset with medical family history criteria. Using these criteria has led to suboptimal case finding (Matloff et al. 2013; Adelson et al. 2014). The threshold was high before an expensive DNA test would be done while only a small proportion of patients would be a mutation carrier. To increase sensitivity and specificity, before DNA testing additional tumour testing was suggested to detect deficiency of mismatch repair proteins by immunohistochemistry (IHC) staining and/or microsatellite instability (MSI) which may be indicative of LS. In addition, a check for identifying sporadic cases in MSI-high tumours to rule out LS could be performed (*MLH1* promoter hypermethylation testing, or *BRAF* V600E mutation testing).

To boost the number of patients identified with LS, in 2009 recommendations were issued by the US CDC Office of Public Health Genomics EGAPP working group (2009) to screen the tumours in all CRC cases via IHC or MSI, and more recently the UK National Institute for Health and Care Excellence (NICE) issued guidance (NICE 2017). Such "universal" tumour screening was found to be cost-effective, though in practice age limits (e.g. all patients under age 70) are observed (Leenen et al. 2016).

Introducing universal tumour screening is not self-evident, even when evidence of effectiveness and cost-effectiveness exists and (inter)national guidelines have been developed. Hospitals are still in the process of adapting their services and currently a range of different pathways for identifying LS patients can be found, some of which still include or rely on the Bethesda criteria rather than using universal tumour screening (Tognetto et al. 2017; Noll et al. 2018).

A lack of awareness of guidelines can hinder implementation (Bellcross et al. 2012). In the United States a guideline exists, but guide familiarity is still suboptimal (Noll et al. 2018). Education of the various stakeholders is highly relevant (Schneider et al. 2006). In Australia Mascarenhas et al. (2018) identified the need for a uniform national testing policy, providing structural guidance. To clarify procedures and referral, local guidelines and protocols need to be developed. Local practices and roles and responsibilities need to be reconsidered for tumour testing to become routine. While previously a medical specialist, most notably the surgeon or gastroenterologist, would request the pathologist to investigate colorectal tumours and referred patients to genetic services, now the pathologist needs to check a sample for MSI/IHC testing irrespective of the assumptions of the specialist. The pathologist becomes a more important stakeholder in the first step to identify index cases with LS (Noll et al. 2018). In a study among health-plan leaders, managers and staff, the importance of the organisational structure supporting the implementation of tumour testing was mentioned as well as the relevance of a department "owning" the screening process (Schneider et al. 2016). West et al. (2017) pointed to "role ambiguity" when interviewing pathologists, oncologists, geneticists, gastroenterologists, and primary care providers. In different phases of the screening process, different disciplines can take the lead, but who does what should be clear. Heald et al. (2013) discussed the importance of interdisciplinary communication for a successful screening pathway and referral. Structural innovation, such as the availability of a multidisciplinary team, could govern such collaboration. The importance of a change agent or "clinical champion" in initiating and governing the process of implementation at a local level has been mentioned (Monahan et al. 2017; Schneider et al. 2016). Such champions may also be important in scaling up local initiatives by informing other hospitals about experiences and best practices. Cost and the availability of resources have been reported as a barrier in implementing tumour screening (Schneider et al. 2016; Bellcross et al. 2012; Noll et al. 2018). This includes lab resources e.g MSI testing and human resources for genetic counselling (Noll et al. 2018) as not all hospitals have such facilities in house. A lack of infrastructure may result in little support related to laboratory processing and patchy systems for electronic ordering and tracking which may hinder routine testing (Schneider et al. 2016). Besides such structural considerations, also more cultural aspects of redefining roles and responsibilities are at stake in routine tumour testing. A question that has been debated is whether consent should be asked for routine tumour testing and whether an opt out option would be relevant (Bombard et al. 2017). In that case the surgeon or gastroenterologist needs to act. Here the different cultures of genetics and mainstream disciplines need to be attuned. In clinical genetics pretest counselling would be the standard, while in mainstream medicine many tests may be done as a routine, and only if an abnormal result is detected, will the patient receive more comprehensive information (Dicks et al. 2019). Especially in oncology increasingly treatment-focused genetic testing calls for clarification of tasks and close collaborations between clinical specialties and clinical geneticists (Douma et al. 2019).

Successful Strategies

For implementation of a genetic test with adequate evidence of clinical utility, various strategies exist (Unim et al. 2019; Mascarenhas et al. 2018). Implementation should not be done haphazardly. Rigorous evaluation could lead to evidence of efficacy and the most cost-effective strategy. Case studies of successful implementation (Cavazza et al. 2019) can lead to a learning curve and aid in scaling up of promising local niche initiatives. After raising the age limit of tumour genetic testing from 50 to 70 years of age, in the Netherlands four times more LS patients were identified when compared to common practice (Sie et al. 2014). Probabilistic efficacy and cost-effectiveness analyses showed that, when family cascade screening was included, the increased age approach was highly cost-effective and more patients could benefit from life-saving surveillance (Sie et al. 2014). Thus, scientific evidence can help to convince stakeholders. In terms of health technology assessment (HTA), it is clear to what extent the intervention is beneficial and a change of practice is required. Public awareness is another relevant factor, as is genetic education of healthcare professionals, preferably building on educational needs expressed by the target group (Houwink et al. 2015) stimulating culture change. Furthermore, statutory registration of genetic specialists, availability of essential staff and equipment, and existence of registries and guidelines allow for structure change and facilitate genetic service development (Rigter et al. 2014).

Familial Hypercholesterolaemia

Case Finding in Primary Care

A high cholesterol increases the risk of cardiovascular events such as myocardial infarction and stroke. Whereas half of the European population currently has elevated cholesterol levels (World Health Organization 2019), about 1 in 200 to 1 in 500 people will have more strongly elevated cholesterol levels because of familial mutations in the LDL receptor gene (most common), apolipoprotein B (Apo B), proprotein convertase subtilisin/kexin 9 (*PCSK9*), or low-density lipoprotein receptor adaptor protein (*LDLRAP*) (Singh and Bittner 2015).

People with FH have a high risk of cardiovascular events that may start already at an early age (Nordestgaard et al. 2013; Singh and Bittner 2015). The use of statins decreases these risks considerably. Various guidelines agree that intensive treatment with lipid-lowering medications should begin as quickly as possible and that lifestyle modifications (not smoking, achieving/maintaining a healthy weight) should be an integral part of the therapy (Nordestgaard et al. 2013; Migliara et al. 2017). Several guidelines advise to start treatment during childhood, requiring DNA testing in children from 8 years of age (Knowles et al. 2017).

Genetic screening of first-degree relatives of index patients followed by the use of statins is a highly cost-effective way to organise FH care (Wonderling et al. 2004; Rosso et al. 2017).

Despite evidence of cost-effectiveness and the availability of national and international professional guidelines, FH is still underdiagnosed (Nordestgaard et al. 2013, Singh and Bittner 2015). Often only parts of the cycle of identification of index patients and tracing first-degree family members are realised in practice. Rarely a screening programme is elaborated and implemented.

The ways in which FH testing and cascade screening of family members is organised differ between countries and regions (Migliara et al. 2017; Aarden et al. 2011). For instance, in the United Kingdom FH care has been organised in mainstream medicine, while in the Netherlands a PH approach was chosen with the implementation of an official cascade screening programme. While the United Kingdom tended to identify and treat patients based on phenotype, in the Netherlands genetic testing formed a central element of FH care since the early 1990s (Aarden et al. 2011). Practices differ with respect to available infrastructure and resources, and (cultural) views on whether DNA tests should be done and by whom, posing different hurdles but also different windows of opportunity for integrating genetic testing.

In a systematic approach one could offer a cholesterol test to all persons attending a health check, a school or all persons working for a specific organisation. Alternatively, opportunistic (phenotypic) screening might occur, for instance during visits to primary care. In principle, the phenotype (high level of LDL and total cholesterol, clinical symptoms, such as tendon xanthomas) in combination with family history (assessing the number of family members with cardiovascular problems below, e.g. age 50 or 60) is the starting point for subsequent investigation and potentially a referral for a DNA test to confirm diagnosis. In primary care practice such initial information is not always available or results are ambiguous contributing to suboptimal case finding and referral. For instance, people with high levels of cholesterol may be referred while not having FH (Weng et al. 2015). Strategies to boost the number of identified cases have included education of primary care physicians (Jackson et al. 2019). An alternative strategy is to make use of automated healthcare information systems.

In the United Kingdom a "Familial Hypercholesterolaemia case ascertainment tool" (FAMCAT) has been developed making use of routine primary care data to enhance detection (Weng et al. 2015). An algorithm integrating nine diagnostic variables showed high discrimination for identifying FH cases. Based on the outcome cases with the highest probability of FH could be selected for further assessment (Weng et al. 2015). This pilot study builds on an existing primary care registration system: Clinical Practice Research Datalink. If the algorithm works in one practice, all practices working with the same standardised codes could use the structured approach of this method to scale up the identification of index cases. Timely identification of index patients allows for adequate treatment. However, crucial in FH care is the fact that diagnosis confirmed by a genetic test allows for cascade screening of first-degree family members who may benefit from primary prevention of cardiovascular problems. In the next paragraph we will discuss a successful example of a systematic way to inform relatives of an index case.

Informing Family Members

In cascade screening for a dominantly inherited condition such as FH parents, brothers, sisters, and children of an index patient are informed and offered testing. Should one of them also have FH, then a new circle of first-degree relatives can be informed (see Fig. 5.1).

In recent decades, discussions addressed how to identify family members and inform them in an ethically responsible way (Newson and Humphries 2005). Studies have shown that only a proportion of first-degree family members are informed of their potential risk, and if they are informed, not all family members make an appointment for genetic counselling and/or testing. Strategies to inform family members may be indirect, e.g. the index patient is requested to inform the family members. This is regarded as an ethically responsible way, as privacy of family members is safeguarded, while they may feel less overwhelmed and pressurised into testing. However, the index may experience this task as a burden, especially when he or she is recently diagnosed or is receiving impactful treatment for the disorder, or family relations are complex or disturbed. The information may not reach the family or may be imprecise, leading to suboptimal uptake. Other approaches involve healthcare professionals contacting family members directly. In some countries this is regarded as an option, which has as advantage that the number of contacted family members may be higher (Newson and Humphries 2005). Currently, indirect approaches prevail, and usually geneticists do not contact family members themselves. There are other options, such as supporting the patient by providing information leaflets and family letters, and discussing during follow-up whether the index patient has succeeded in informing family members. While clinical geneticists can be regarded as specialists in dealing with familial information, in the case of FH, however, patients are often not referred to the genetics department. FH patients are often diagnosed by specialists in internal or vascular medicine, because the genetics is regarded as relatively straightforward. FH is regarded as a treatable disorder for which expensive counselling is not always seen as self-evident. Consequently, in many countries the procedure for cascade screening also is in the hands of a non-genetic specialty, lacking skills and resources to address challenges of stimulating familial investigations.

In addition to suboptimal case finding, also patchy mechanisms for cascade screening contribute to the fact that, internationally, FH is still underdiagnosed. PH initiatives have been developed in several countries to organise cascade testing in a more systematic way. A well-known example is the national screening programme for FH in the Netherlands. This programme was coordinated by the Foundation for Tracing FH (StOEH). After the identification of index patients in primary care and vascular medicine, the index patient was subsequently asked to provide contact details of first-degree family members to the screening organisation. The index patient would have asked his family members if they agreed to share their details. Family members were contacted by the screening organisation and visited at home or in the workplace. During family gatherings information would be given by genetic field workers who could draw blood for DNA testing at the same event.

A national registry was updated with the mutation details of each family, so that whenever a new patient was found it could be checked whether this case could be connected to existing entries in the database. In that case relatively cheap confirmatory testing of the known familial mutation could be performed. While the identification of index cases was part of regular healthcare, the tracing, informing, and diagnosis of family members was funded by PH budgets.

The screening programme was very successful. At its peak on average eight family members of an index patient were tested (Knowles et al. 2017).

After 20 years the programme came to an end in 2014, as it had always been regarded as a project. Initially it was thought that most mutation carriers would have been found by that time, but as more precise prevalence data became available it turned out that not 1 in 500 but even up to 1 in 200 might be affected. One of the arguments not to extend the programme was the fact that the active and direct approach to visit family members at home would run counter to recent trends to respect privacy and autonomy. Healthy family members of FH patients should instead contact healthcare themselves after being informed by the patient. The ending of the programme posed a challenge to healthcare professionals as they were expected to take up the tasks previously coordinated by the screening organisation. After ending the programme, the number of family screenings dropped from over 2000 yearly to around 400 in 2015 (Louter et al. 2017). While the routines to identify index cases had not changed, now no systematic effort was made to organise cascade screening. A new coordinating organisation was erected, Dutch Expertise Centre for Inheritance Testing of Cardiovascular Diseases (LEEFH) that received only a very modest budget and was faced with the task to mobilise stakeholders. Index patients were contacted by LEEFH and given family letters to distribute. Regional differences in the percentage of patients detected had always existed. These can be further explored to see which local and regional initiatives contain successful strategies for case finding and cascade testing. For instance, some hospital departments use nurses to discuss informing family members with patients. Some local initiatives stimulate education for primary care physicians about FH and the relevance of informing the family, and discuss what role primary care physicians can take, as it is not self-evident that they feel responsible for family members of their patients. Other initiatives include experimenting with lab alerts for physicians in case of high cholesterol and other relevant markers allowing for more structural support (Broen et al. 2016; Louter et al. 2017; Van El et al. 2018).

In the background of these developments in practice, internationally in clinical genetics a discussion is ongoing on pros and cons of the indirect approach. In recent lawsuits the right to be informed has been stressed, arguing for more than a moral duty to inform family members in case of preventable serious health damage (Parker and Lucassen 2018). In the Netherlands, a new multidisciplinary guideline has been drafted to grant the medical specialist more leeway to contact family members directly in case the index patient needs more support (Federatie Medische Specialisten 2019). Such guideline development can create a new window of opportunity to help increase the number of family members that are being informed by creating a new *structure* and introducing a novel *culture*. However, it is not clear yet to what extent non-genetic professionals will feel comfortable using such a guideline, and whether primary care physicians will be influenced by such new standards. Alternatively, new information and communication technology may help in changing the necessary infrastructure to aid informing family members and offering easily accessible information via websites and email messages.

5.3 Discussion

Martina C. Cornel, Tessel Rigter, and Carla G. van El

We have discussed several technical developments that make a shift from cure to prevention possible for families with monogenic subtypes of hereditary cancers and CVDs. The available knowledge and techniques tend to be implemented sub-optimally. The examples from LS and FH show that systematic approaches can be highly efficient to stimulate finding index patients and inform healthy family members of their risk. A dedicated effort is needed to implement and optimise subsequent elements of the cycle of identification of index patients and tracing first-degree family members. Systematic case finding and cascade testing can only be successful tools if they are made to fit and adapt healthcare practices as increasingly non-genetic healthcare professionals are confronted with monogenic subtypes of common disorders. Awareness and education of healthcare professionals is relevant, while coordination and attunement between primary and secondary care and between genetic and non-genetic healthcare professionals must be organised. Roles and responsibilities have to be redistributed and redefined, and require new ways of doing (practice), thinking (culture) and organising (structure). Following ongoing developments in technology, infrastructure, demand, and acceptability, windows of opportunity may occur that allow scaling up of local initiatives. Governance is needed that may involve change agents, supported by healthcare policy, professional organisations and governments.

The different technologies that will be of relevance in the next decade include genomic information as well as data infrastructures (making data Findable, Accessible, Interoperable, Reusable) (Wilkinson et al. 2016). True multidisciplinary efforts are needed for public and professional education, the development of guidelines, HTA, and systematic evaluation of health gain achieved. These efforts need to be planned and resources need to be made available. A dedicated effort is necessary to stimulate further responsible implementation of evidence-based interventions in healthcare to inform family members in cases of hereditary subtypes of common diseases, such as hereditary cancers or CVDs.

Acknowledgements The authors thank Valentina Baccolini, Renate van den Broek, Peter Piko and Matthijs Loning for their contributions to this part of the PRECeDI project.

References

Aarden, E., Van Hoyweghen, I., & Horstman, K. (2011). The paradox of public health genomics: definition and diagnosis of familial hypercholesterolaemia in three European countries. *Scandinavian Journal of Public Health, 39*, 634–639.

Achterbergh, R., Lakeman, P., Stemerding, D., et al. (2007). Implementation of preconceptional carrier screening for cystic fibrosis and haemoglobinopathies: a sociotechnical analysis. *Health Policy, 83*, 277–286.

Adelson, M., Pannick, S., East, J. E., Risby, P., et al. (2014). UK colorectal cancer patients are inadequately assessed for Lynch syndrome. *Frontline Gastroenterology, 5*, 31–35.

Andermann, A., Blancquaert, I., Beauchamp, S., et al. (2008). Revisiting Wilson and Jungner in the genomic age: a review of screening criteria over the past 40 years. *Bulletin of the World Health Organization, 86*, 317–319.

Bellcross, C. A., Bedrosian, S. R., Daniels, E., et al. (2012). Implementing screening for Lynch syndrome among patients with newly diagnosed colorectal cancer: summary of a public health/clinical collaborative meeting. *Genetics in Medicine: Official Journal of the American College of Medical Genetics, 14*, 152–162.

Bombard, Y., Rozmovits, L., Sorvari, A., et al. (2017). Universal tumor screening for Lynch syndrome: healthcare providers' perspectives. *Genetics in Medicine, 19*, 568–574.

Broen, K., Gidding, L., Houter, M., et al. (2016). Op zoek naar familiaire hypercholesterolemie (Looking for FH). *Medisch Contact, 16*, 20–23.

Brosco, J. P., & Paul, D. B. (2013). The political history of PKU: Reflections on 50 years of newborn screening. *Pediatrics, 132*, 987–989.

Cavazza, A., Radia, C., Harlow, C., et al. (2019). Experience of the implementation and outcomes of universal testing for Lynch syndrome in the United Kingdom. *Colorectal Disease, 21*, 760–766.

Dicks, E., Pullman, D., Kao, K., et al. (2019). Universal tumor screening for Lynch syndrome: Perspectives of Canadian pathologists and genetic counselors. *Journal of Community Genetics, 10*, 335–344.

Douma, K. F. L., Bleeker, F. E., Medendorp, N. M., et al. (2019). Information exchange between patients with Lynch syndrome and their genetic and non-genetic health professionals: Whose responsibility? *Journal of Community Genetics, 10*, 237–247.

Essink, D. R. (2012). *Sustainable health systems: The role of change agents in health system innovation*. Amsterdam: Dissertation VU University.

Evaluation of Genomic Applications in Practice and Prevention (EGAPP) Working Group. (2009). Recommendations from the EGAPP Working Group: genetic testing strategies in newly diagnosed individuals with colorectal cancer aimed at reducing morbidity and mortality from Lynch syndrome in relatives. *Genetics in Medicine, 11*, 35–41.

Federatie Medische Specialisten. (2019). Retrieved from https://richtlijnendatabase.nl/richtlijn/informeren_van_familieleden_bij_erfelijke_aandoeningen/startpagina_-_informeren_van_familieleden_bij_erfelijke_aandoeningen.html. Accessed 10 Oct 2019.

Gabai-Kapara, E., Lahad, A., Kaufman, B., et al. (2014). Population-based screening for breast and ovarian cancer risk due to *BRCA1* and *BRCA2*. *Proceedings of the National Academy of Sciences of the United States of America, 111*, 14205–14210.

Geels, F. W. (2002). Technological transitions as evolutionary reconfiguration processes: A multilevel perspective and a case-study. *Research Policy, 31*, 1257–1274.

Geels, F. W., & Schot, J. (2007). Typology of sociotechnical transition pathways. *Research Policy, 36*, 399–417.

Henneman, L., Van El, C. G., & Cornel, M. C. (2013). Genetic testing and implications for personalized medicine: Changes in public and healthcare professional perspectives. *Personalized Medicine, 10*, 217–219.

Houwink, E. J., Muijtjens, A. M., van Teeffelen, S. R., et al. (2015). Effect of comprehensive oncogenetics training interventions for general practitioners, evaluated at multiple performance levels. *PLoS One, 10*, e0122648.

Jackson, L., O'Connor, A., Paneque, M., et al. (2019). The Gen-Equip Project - evaluation and impact of genetics e-learning resources for primary care in six European languages. *Genetics in Medicine, 21*, 718–726.

Khoury, M. J., Berg, A., Coates, R., et al. (2008). The evidence dilemma in genomic medicine. *Health Affairs (Millwood), 27*, 1600–1611.

Khoury, M. J., Gwinn, M., Yoon, P. W., et al. (2007). The continuum of translation research in genomic medicine: How can we accelerate the appropriate integration of human genome discoveries into health care and disease prevention? *Genetics in Medicine, 9*, 665–674.

Knowles, J. W., Rader, D. J., & Khoury, M. J. (2017). Cascade screening for familial hypercholesterolemia and the use of genetic testing. *Journal of the American Medical Association, 318*, 381–382.

Leenen, C. H., Goverde, A., de Bekker-Grob, E. W., et al. (2016). Cost-effectiveness of routine screening for Lynch syndrome in colorectal cancer patients up to 70 years of age. *Genetics in Medicine, 18*, 966–973.

Louter, L., Defesche, J., & Roeters van Lennep, J. (2017). Cascade screening for familial hypercholesterolemia: Practical consequences. *Atherosclerosis. Supplements, 30*, 77–85.

Mascarenhas, L., Shanley, S., Mitchell, G., et al. (2018). Current mismatch repair deficiency tumor testing practices and capabilities: A survey of Australian pathology providers. *Asia-Pacific Journal of Clinical Oncology, 14*, 417–425.

Matloff, J., Lucas, A., Polydorides, A. D., et al. (2013). Molecular tumor testing for Lynch syndrome in patients with colorectal cancer. *Journal of the National Comprehensive Cancer Network, 11*, 1380–1385.

Migliara, G., Baccolini, V., Rosso, A., et al. (2017). Familial hypercholesterolemia: A systematic review of guidelines on genetic testing and patient management. *Frontiers in Public Health, 5*, 252.

Monahan, K. J., Alsina, D., Bach, S., et al. (2017). Urgent improvements needed to diagnose and manage Lynch syndrome. *BMJ, 356*, j1388.

Ned, R. M., & Sijbrands, E. J. (2011). Cascade screening for familial hypercholesterolemia (FH). *PLoS Currents, 3*, RRN1238.

Newson, A. J., & Humphries, S. E. (2005). Cascade testing in familial hypercholesterolaemia: How should family members be contacted? *European Journal of Human Genetics, 13*, 401–408.

NICE. (2017). Molecular testing strategies for Lynch syndrome in people with colorectal cancer. Retrieved from https://www.nice.org.uk/guidance/DG27/chapter/3-The-diagnostic-tests. Accessed 16 Oct 2019.

Noll, A., Parekh, P. J., Zhou, M., et al. (2018). Barriers to Lynch syndrome testing and preoperative result availability in early-onset colorectal cancer: A national physician survey study. *Clinical and Translational Gastroenterology, 9*, 185.

Nordestgaard, B. G., Chapman, M. J., Humphries, S. E., et al. (2013). Familial hypercholesterolaemia is underdiagnosed and undertreated in the general population: guidance for clinicians to prevent coronary heart disease: consensus statement of the European Atherosclerosis Society. *European Heart Journal, 34*, 3478–3490.

Parker, M., & Lucassen, A. (2018). Using a genetic test result in the care of family members: How does the duty of confidentiality apply? *European Journal of Human Genetics, 26*, 955–959.

Pitini, E., De Vito, C., Marzuillo, C., et al. (2018). How is genetic testing evaluated? A systematic review of the literature. *European Journal of Human Genetics, 26*, 605–615.

Rigter, T., Henneman, L., Broerse, J. E., et al. (2014). Developing a framework for implementation of genetic services: Learning from examples of testing for monogenic forms of common diseases. *Journal of Community Genetics, 5*, 337–347.

Rigter, T., Jansen, M. E., de Groot, J. M., et al. (2020). Implementation of pharmacogenetics in primary care: A multi-stakeholder perspective. *Frontiers in Genetics, 11*, 10. eCollection 2020. https://doi.org/10.3389/fgene.2020.00010.

Roa, B. B., Boyd, A. A., Volcik, K., & Richards, C. S. (1996). Ashkenazi Jewish population frequencies for common mutations in *BRCA1* and *BRCA2*. *Nature Genetics, 14*, 185–187.

Rosso, A., Pitini, E., D'Andrea, E., et al. (2017). The cost-effectiveness of genetic screening for familial hypercholesterolemia: A systematic review. *Annali di Igiene, 29*, 464–480.

Rotmans, J., Kemp, R., & van Asselt, M., et al. (2001). Transitions & transition management: The case for low emission energy supply. (ICIS Working Paper; Vol. I01-E001). Maastricht: ICIS.

Roberts, M., Kennedy, A., Chambers, D. et al. (2017). The current state of implementation science in genomic medicine: opportunities for improvement. *Genetics in Medicine: Official Journal of the American College of Medical Genetics, 19*, 858–863.

Schneider, J. L., Davis, J., Kauffman, T. L., et al. (2016). Stakeholder perspectives on implementing a universal Lynch syndrome screening program: A qualitative study of early barriers and facilitators. *Genetics in Medicine, 18*, 152–161.

Severin, F., Borry, P., Cornel, M. C., et al. (2015). Points to consider for prioritizing clinical genetic testing services: A European consensus process oriented at accountability for reasonableness. *European Journal of Human Genetics, 23*, 729–735.

Sie, A. S., Mensenkamp, A. R., Adang, E. M., et al. (2014). Fourfold increased detection of Lynch syndrome by raising age limit for tumour genetic testing from 50 to 70 years is cost-effective. *Annals of Oncology, 25*, 2001–2007.

Singh, S., & Bittner, V. (2015). Familial hypercholesterolemia—Epidemiology, diagnosis, and screening. *Current Atherosclerosis Reports, 17*, 482.

Tognetto, A., Michelazzo, M. B., Calabró, G. E., et al. (2017). A systematic review on the existing screening pathways for Lynch syndrome identification. *Frontiers in Public Health, 5*, 243.

Unim, B., Pitini, E., Lagerberg, T., et al. (2019). Current genetic service delivery models for the provision of genetic testing in Europe: A systematic review of the literature. *Frontiers in Genetics, 10*, 552.

van den Bosch, S. (2010). Transition experiments: exploring societal changes towards sustainability. Dissertation, Chapter 3, Erasmus University Rotterdam.

van El, C. G., Baccolini, V., Piko, P., et al. (2018). Stakeholder views on active cascade screening for familial hypercholesterolemia. *Healthcare (Basel), 6*(3), E108.

van El, C. G., & Cornel, M. C. (2011). Genetic testing and common disorders in a public health framework. *European Journal of Human Genetics, 19*, 377–381.

van Raak, R. (2010). The transition (management) perspective on long-term change in healthcare. In J. E. W. Broerse & J. F. G. Bunders (Eds.), *Transitions in health systems: Dealing with persistent problems* (pp. 49–86). Amsterdam: VU University Press.

Weng, S. F., Kai, J., Andrew Neil, H., et al. (2015). Improving identification of familial hypercholesterolaemia in primary care: Derivation and validation of the familial hypercholesterolaemia case ascertainment tool (FAMCAT). *Atherosclerosis, 238*, 336–343.

West, K. M., Burke, W., & Korngiebel, D. M. (2017). Identifying "ownership" through role descriptions to support implementing universal colorectal cancer tumor screening for Lynch syndrome. *Genetics in Medicine: Official Journal of the American College of Medical Genetics, 19*, 1236–1244.

Wilkinson, M., Dumontier, M., Aalbersberg, I., et al. (2016). The FAIR Guiding Principles for scientific data management and stewardship. *Science Data, 3*.

Wonderling, D., Umans-Eckenhausen, M. A., Marks, D., et al. (2004). Cost-effectiveness analysis of the genetic screening program for familial hypercholesterolemia in The Netherlands. *Seminars in Vascular Medicine, 4*, 97–104.

World Health Organization. (2019). Global Health Observatory data. Retrieved from http://apps.who.int/gho/data/view.main.2570?lang=en. Accessed 10 Oct 2019.

Chapter 6
Identification of Organisational Models for the Provision of Predictive Genomic Applications

Corrado De Vito, Brigid Unim, Martina C. Cornel, Anant Jani, Muir Gray, and Jim Roldan

6.1 Delivery Models for Genetic/Genomic Applications

Brigid Unim

A considerable number of personalised medicine (PM) initiatives have been launched worldwide since the completion of the first sequence of the human genome in 2003 (Collins et al. 2003). This has led to the rapid diffusion of genetic tests for common diseases offered in the public and private sectors. Despite the fast and promising development of genomic applications, guaranteeing quality standards of genetic services poses serious concerns considering the multilevel phases of genetic service delivery. The critical elements in the pre-testing phase are the lack or insuf-

C. De Vito (✉) · B. Unim
Department of Public Health and Infectious Diseases, Sapienza University of Rome, Rome, Italy
e-mail: corrado.devito@uniroma1.it; brigid.unim@uniroma1.it

M. C. Cornel
Amsterdam University Medical Centers, location VUMC, Vrije Universiteit Amsterdam, Department of Clinical Genetics, Section Community Genetics, and Amsterdam Public Health Research Institute, Amsterdam, The Netherlands
e-mail: mc.cornel@amsterdamumc.nl

A. Jani · M. Gray
Value Based Healthcare Programme, Department of Primary Care, University of Oxford, Oxford, UK
e-mail: anant.jani@phc.ox.ac.uk; muir.gray@phc.ox.ac.uk

J. Roldan
Linkcare Health Services, Barcelona, Spain
e-mail: jimroldan@linkcareapp.com

© The Editor(s) and Author(s), under exclusive license to Springer Nature Switzerland AG 2021
S. Boccia et al. (eds.), *Personalised Health Care*, SpringerBriefs in Public Health, https://doi.org/10.1007/978-3-030-52399-2_6

ficient data on the validity or utility of numerous genetic applications already introduced in practice (Khoury et al. 2007), the qualification of healthcare professionals providing genetic counselling and what constitutes an appropriate consent. In the testing phase, laboratory standards and the interface between laboratory and clinical services are equally important to the process of ensuring quality. In the post-test phase, the results of the test must be interpreted by qualified healthcare professionals who can appropriately counsel the patient regarding available interventions and existing support structures. Above all, some countries have not enacted specific regulations governing the use of genetic applications in clinical and public health (PH) practice. For instance, the additional protocol to the "Convention for the Protection of Human Rights and Dignity of the Human Being with regard to the Application of Biology and Medicine" states that "a genetic test for health purposes may only be performed under individualised medical supervision" in the European Union (EU). The protocol has been signed by 10 EU member states but only ratified by six of them (i.e. Moldova, Montenegro, Norway, Slovenia, Portugal and Czech Republic) (Council of Europe 2019). In Canada, there are few regulatory frameworks that are directly relevant to genetic tests, services or programmes. Services provided by healthcare professionals are governed by pre-existing common law and civil law norms. Moreover, there is no regulatory framework for direct-to-consumer genetic testing (DTC-GTs) (Little et al. 2009).

These concerns contribute to the lack of evidence on what constitutes an optimal genetic service delivery model, defined as the broad context within the Public Health Genomics (PHG) framework in which genetic services are offered to individuals and families with or at risk of genetic disorders. A genetic service delivery model is a combination of personal healthcare services provided by healthcare professionals to individuals and families (i.e. diagnosis, treatment/management and information) and PH services and functions (i.e. population screening, financing, policy development, workforce education, information/citizen empowerment, service evaluation and research) (Unim et al. 2017).

Current Organisations of Genetic Services

Each genetic programme (i.e. healthcare programmes providing a genetic test) is defined by unique factors, such as: (i) practice setting and financial resources (public vs. private); (ii) service provider and patient access [geneticists vs. primary care physicians/other medical specialists (e.g. cardiologists, oncologists, neurologists and so on)]; (iii) policy regulation (national and local policies, guidelines, protocols and position statements); (iv) laboratory practice standards (quality control standards, qualified personnel, etc.) and (v) information dissemination (methods of providing information about genetic services to patients and service providers) (Washington State Department of Health 2008). The analysis of the genetic programmes identified in literature records, according to the aforementioned unique factors, laid the basis for the classification of genetic service delivery models in

Europe and in selected extra European countries (i.e. the USA, Canada, Australia and New Zealand). Current genetic service delivery models for the provision of genetic testing are classified into five categories (Table 6.1) according to the health-care professional with the most prominent role in genetic test provision, treatment and monitoring of patients and the care pathways (i.e. a patient's path through different healthcare professionals from the initial point of access to healthcare services to treatment of the genetic disorder and follow-up) (Unim et al. 2019). A detailed description of the models is reported below.

Model I: Genetic services led by geneticists. The genetic team may include medical geneticists, genetic counsellors and other healthcare workers (e.g. genetic nurses). The genetic team is responsible for risk assessment, counselling and testing of individuals or families affected by or at risk for genetic disorders. Depending on the case, the team collaborates with other medical specialists (e.g. oncologists, cardiologists, nephrologists) who could be part of the genetic service (e.g. multidisciplinary genetic clinics). The access of patients to this model of genetic services may occur through two different pathways:

(a) Patient → GP or Medical specialist → Counsellor → Lab;
(b) Patient → Counsellor → Lab.

Table 6.1 Genetic service Delivery Models according to the roles of the healthcare professionals involved in patients' pathways to care

Pathway	Model I: Genetic services led by geneticists	Model II: Primary care model	Model III: Medical specialist model	Model IV: Genetic services integrated into population screening programmes	Model V: DTC model
I	Patient → GP or Medical specialist → Counselor → Lab	Patient → GP → Counselor → Lab	Patient → (GP) Medical specialist → Lab	Patient → GP or Medical specialist → Counselor → Lab	Patient → Lab
II	Patient → Counselor → Lab	Patient → GP → Lab	Patient → (GP) Medical specialist → Counselor → Lab	Patient → GP or Medical specialist → Lab	
III				Patient → Counselor→ Lab	

GP general practitioner, *DTC* direct-to-consumer, *Counsellor* counselling could be provided by medical geneticists or genetic counsellors
Source: Reproduced from Unim B, Pitini E, Lagerberg T, et al. (2019) Current Genetic Service Delivery Models for the Provision of Genetic Testing in Europe: A Systematic Review of the Literature. Front. Genet. 10:552, Table 4. Some modifications to the text were made. https://doi.org/10.3389/fgene.2019.00552, licensed under the terms of the Creative Commons Attribution License (https://creativecommons.org/licenses/by/4.0/)

The first pathway (Ia) occurs when a patient seeks medical assistance from a General Practitioner (GP) or any specialist doctor who makes a referral to the genetic service where a genetic counsellor or a medical geneticist can perform a risk assessment and may suggest genetic testing to the patient. Based on results of the test, genetic counsellors or medical geneticists may suggest surveillance recommendation and/or management intervention. The clinical management of genetic conditions may involve different medical specialists, other than geneticists (e.g. oncologists, cardiologists, nephrologists, endocrinologists). The second pathway (Ib) occurs when a patient, without a GP or medical specialist referral, contacts the genetic service where a genetic counsellor or a medical geneticist can perform a risk assessment. Pathway Ib corresponds to pathway Ia from this point upward. Model I has been identified in the United Kingdom, the United States and Australia. The main genetic tests offered under Model I are BRCA1/2, Lynch syndrome (LS) and newborn screening panel (Fig. 6.1a) (Unim et al. 2019).

Model II: Primary care model. A prominent role is played by primary care units, which may include primary care physicians (GPs or family physicians), a nurse practitioner or a physician assistant. In these units, GPs have some training in genetics and can undertake an initial risk assessment using standardised referral guidelines. GPs may be able to manage patients with or at risk of genetic disorders without consulting medical geneticists or genetic counsellors. The pathways associated to this model are:

(a) Patient→ GP → Counsellor →Lab
(b) Patient → GP → Lab

Pathway IIa occurs when a patient contacts a GP who undertakes a risk assessment and then makes referrals to a genetic service, where a genetic counsellor or a medical geneticist can perform counselling and suggest genetic testing to the patient. Pathway IIb occurs when a patient contacts a GP who can perform a risk assessment, undertake counselling and suggest genetic testing. Model II is prevalent in the United Kingdom and the United States. The main genetic tests offered under Model II are BRCA1/2, LS, familial hypercholesterolaemia (FH) and diabetes (Fig. 6.1b) (Unim et al. 2019).

Model III: Medical specialist model. Genetic tests can be requested directly by medical specialists (e.g. oncologists, cardiologists, neurologists) who may be able to manage patients with or at risk of genetic disorders without consulting medical geneticists or genetic counsellors. The possible pathways in Model III are:

(a) Patient → (GP) Medical specialist → Lab
(b) Patient→ (GP) Medical specialist→ Counsellor→ Lab

Pathway IIIa occurs when a patient contacts (with or without a GP referral) a medical specialist who performs a risk assessment, undertakes genetic counselling and suggests genetic testing. In pathway IIIb, a patient contacts (with or without a GP referral) a medical specialist who undertakes the initial risk assessment and then requests counselling, collaborating with the medical geneticist or genetic counsellor in the management of the patient. Model III is prevalent in the United Kingdom, the United States, Australia and France. The main genetic tests offered under this model are BRCA1/2, LS and FH (Fig. 6.1c) (Unim et al. 2019).

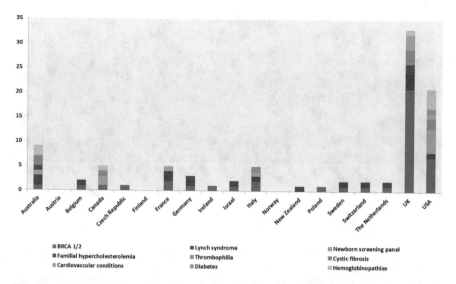

Fig. 6.1 (**a**) Geographic distribution of genetic tests according to genetic service delivery models, genetic services led by geneticists. Reprinted from Unim B, Pitini E, Lagerberg T, et al. (2019) Current Genetic Service Delivery Models for the Provision of Genetic Testing in Europe: A Systematic Review of the Literature. Front. Genet. 10:552, Fig. 3. https://doi.org/10.3389/fgene.2019.00552, licensed under the terms of the Creative Commons Attribution License (https://creativecommons.org/licenses/by/4.0/)

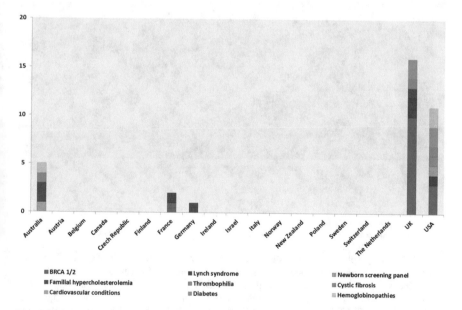

Fig. 6.1 (**b**) Geographic distribution of genetic tests according to genetic service delivery models, the primary care model. Reprinted from Unim B, Pitini E, Lagerberg T, et al. (2019) Current Genetic Service Delivery Models for the Provision of Genetic Testing in Europe: A Systematic Review of the Literature. Front. Genet. 10:552, Fig. 4. https://doi.org/10.3389/fgene.2019.00552, licensed under the terms of the Creative Commons Attribution License (https://creativecommons.org/licenses/by/4.0/).

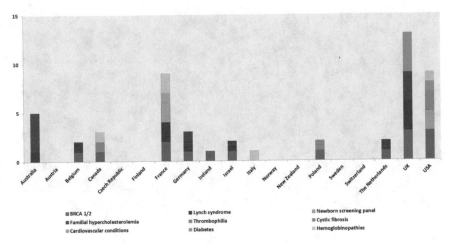

Fig. 6.1 (**c**) Geographic distribution of genetic tests according to genetic service delivery models, the medical specialist model. Reprinted from Unim B, Pitini E, Lagerberg T, et al. (2019) Current Genetic Service Delivery Models for the Provision of Genetic Testing in Europe: A Systematic Review of the Literature. Front. Genet. 10:552, Fig. 5. https://doi.org/10.3389/fgene.2019.00552, licensed under the terms of the Creative Commons Attribution License (https://creativecommons.org/licenses/by/4.0/)

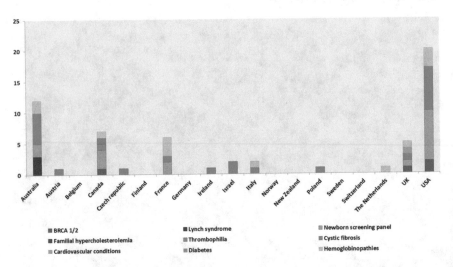

Fig. 6.1 (**d**) Geographic distribution of genetic tests according to genetic service delivery models, genetic services integrated into population screening programmes. Reprinted from Unim B, Pitini E, Lagerberg T, et al. (2019) Current Genetic Service Delivery Models for the Provision of Genetic Testing in Europe: A Systematic Review of the Literature. Front. Genet. 10:552, Fig. 6. https://doi.org/10.3389/fgene.2019.00552, licensed under the terms of the Creative Commons Attribution License (https://creativecommons.org/licenses/by/4.0/)

Model IV: Genetic services integrated into population screening programmes. In this model, genetic services are provided within organised population screening programmes, such as Hereditary Breast and Ovarian Cancer (HBOC) screening, colorectal cancer (CRC) screening and the population-based screening of Ashkenazi Jews. There are three possible patient pathways in Model IV:

(a) Patient → GP or Medical specialist → Counsellor → Lab;
(b) Patient → GP or Medical specialist → Lab;
(c) Patient → Counsellor → Lab.

The IVa pathway occurs when a patient takes part in a population-based screening programme; a healthcare professional involved in the screening programme can perform an initial risk assessment and refer the patient to genetic counselling. The genetic counsellor or medical geneticist can provide advice, suggest genetic testing and, based on results of the test, can make recommendations as to monitoring and/ or clinical intervention. In pathway IVb, healthcare professionals involved in a population-based screening programme can perform risk assessment, provide advice and suggest genetic testing. Based on results of the test, the healthcare professional can make recommendations as to monitoring and/or clinical intervention. In pathway IVc, a patient contacts a genetic counsellor or a medical geneticist who can provide advice, suggest genetic testing and, based on results of the test, can recommend monitoring through available population-based screening programmes and/or clinical intervention. Model IV is common in the United States, Australia and the United Kingdom. The main genetic tests offered under Model IV are cystic fibrosis, newborn screening panel and haemoglobinopathies screening (Fig. 6.1d) (Unim et al. 2019).

Model V: Direct-to-consumer (DTC) model. Private companies offer genetic services to consumers, typically through websites. Healthcare professionals are usually not involved in the process and medical referrals are usually not required for genetic testing. The main pathway associated with Model V is Patient → Lab. Model V was identified in the United Kingdom, the United States and New Zealand (Unim et al. 2019). Genetic tests for various applications, including health, pharmacogenomics, ancestry, fitness, nutrition and more, are offered under this model (Phillips 2016).

Genetic services led by geneticists is the most common model of delivery and corresponds to the *classic* model of genetic services (e.g. for rare diseases) provided mainly by geneticists. In recent years, genetic applications are increasingly utilised by other medical specialists who are involved to various degrees in patient management and can also assume a prominent role in patient care according to the underlying genetic disorder. This is indisputable in genetic services led by medical specialists, which is the second most common model of delivery. For instance, oncologists have the most prominent role in genetic test provision and the monitoring of patients undergoing BRCA1/2 or LS testing, while GPs and cardiologists have a prominent role in the case of familial thrombophilia testing. Genetic services are also progressively integrated into population-based screening programmes. This is a relatively new model of delivery and not yet widely implemented. The primary care model is currently one of the least represented models of delivery. It has been outweighed by the medical specialist model due to lack of or limited knowledge of primary care physicians in genetics. The appropriate model for genetic service provision in a specific setting could be defined taking into consideration the main components of a genetic programme (i.e. target population, genetic counselling, genetic testing, diagnosis of carrier status, the healthcare pathway based on the carrier status) and the healthcare system (Unim et al. 2019).

The number of DTC companies and the demand for their services have increased dramatically in recent years. It is likely that the use of these services will continue to grow given the easy access to genetic testing offered by commercial companies, and the increasing tendency to purchase products through the internet (Phillips 2016; Unim et al. 2019). DTC services raise many ethical issues, as testing is not being conducted in a medical setting which would better support the consumer in understanding and managing the decision to undertake testing and the testing results. The rapid development of genetic applications should be therefore accompanied by appropriate regulations that can ensure quality and safety of genetic testing and related services.

Barriers to the Appropriate Implementation of Genetic Discoveries

Most genetic tests have considerable evidence of efficacy and cost-effectiveness and could be fully implemented in clinical and public health practice (i.e. BRCA1/2, LS, FH). However, not all genetic tests have been evaluated prior to their introduction in routine practice (e.g. cardiovascular conditions, type 2 diabetes (T2D), hereditary hemochromatosis). This highlights the need for evidence-based technology assessments and economic evaluations prior to the introduction of genomics applications in mainstream medicine. Only genetic tests with considerable effectiveness and cost-effectiveness data should be offered to citizens as a right to benefit from innovative healthcare (Unim et al. 2019). The lack of national policies governing the use of genomic applications, the lack of expertise in medical genetics (Washington State Department of Health 2008) and funding for genomic research (Pohlhaus and Cook-Deegan 2008) are also barriers to the successful implementation of genomic discoveries. In light of this, professional education in genomics medicine, adequate funding, public policies, and public awareness should be enhanced as the key to the successful implementation of genomic applications in clinical and PH practice.

6.2 Barriers and Facilitating Factors for Implementation of Genetic Services

Martina C. Cornel

Although genomics knowledge has developed very fast in the recent decades, applications in healthcare are still modest, especially for predictive genomic applications. When we speak about healthcare, in fact we often refer to the care for people who come to the hospital because of illness. Healthcare for people without health problems typically is organised in PH services, a completely different silo. Organisational models for ill people and healthy people differ. PH physicians typically give lifestyle advice, fight smoking and organise vaccination programmes for healthy people. The involvement of PH in genetic services is limited, mainly to population screening of newborns (Unim et al. 2019). The new predictive genomic technologies require a change of culture in healthcare towards taking care of health instead of taking care of illness. In terms of structure this implies that different programmes are needed: identifying people at high risk and offer interventions before they fall ill. Not only DNA technologies will be needed in these preventive services, but also traditional skills such as taking a family history and ICT services to combine risk estimates from different sources in algorithms. Predictive genomics also implies that different actors need to take responsibility for genetic services, for instance PH workers instead of clinical geneticists (see Chap. 4, Sect. 4.2.3 for a theoretical model of the constellation perspective and network of actors; Rigter et al. 2014). The organisation of a screening programme often is executed in PH, while DNA technologies typically are used in Centres for Clinical Genetics.

Breast Cancer in Predictive Genomic Applications

In many countries public health organisations are involved in breast cancer screening. Most countries offer biennial mammography to women at 50–74 years of age, which reduces mortality by ±20% (Shah and Guraya 2017). For younger women, relatively more false positives occur, and disadvantages of biennial mammography screening on average outweigh advantages. Screening at a younger age may be worthwhile for women at higher risk, such as first-degree relatives of breast cancer patients. The group of women at the highest risk of breast cancer are women carrying a mutation in the breast cancer genes *BRCA1* or *BRCA2*. The abbreviation BRCA stands for BReast CAncer. The lifetime risk of breast cancer for women carrying a pathogenic *BRCA* mutation is 60–80%, while also their risk of ovarian cancer is increased. Nowadays we know that many more genes are involved in breast cancer risk, such as *ATM, CDH1, CHEK2, PALB2* and *TP53* (La Duca et al. 2020). One of the challenges is obviously to identify women carrying pathogenic mutations in one of these genes as soon as possible when they are adults.

The identification of these healthy women at high risk starts from index patients with cancer or from people asking questions about their family history. In many countries they can be referred to Clinical Genetic Centres when referral criteria are met. DNA testing can be ordered in these centres, when indicated and consented. Some of the healthy women identified as carriers of a high risk may choose prophylactic mastectomy, others may follow surveillance to identify any tumour as soon as possible, to improve their long-term outcomes. Surveillance for women with a higher risk is often not organised within public health organisations, but within specialised clinics. The identification of index cases with mutations in high-risk genes depends on oncologists (for cases affected with cancer) and clinical geneticists (for cases and relatives). Often this is done on a case-by-case basis. At least in theory there are opportunities to improve the identification of women at high risk for hereditary cancer predisposition (La Duca et al. 2020), including modified testing criteria. If for instance age at diagnosis constraints would be relaxed, a larger number of index cases with pathogenic mutations would be recognised. Furthermore, the authors suggest to consider using multigene panel testing for patients with a wide spectrum of cancer histories.

Apart from these high-risk genes, recent studies discussed polygenic risk scores, where hundreds of single-nucleotide polymorphisms (SNPs) are combined in a score, which could be used to stratify women according to breast cancer risk (Mavaddat et al. 2019). The risk distribution for the lowest vs. highest percentiles varied from a few percent to over 30%. Earlier, Chatterjee et al. (2016) showed how polygenic risk scores, including both high-risk genes and low-risk genes, could help to determine the age at which the risk of developing breast cancer reaches a threshold of 2.4% within the next 10 years. At that age surveillance should begin. The risk level of 2.4% corresponds to the average population 10-year risk of developing breast cancer for women at the currently recommended starting age for screening in the United Kingdom. For 20% of women this threshold would be reached at 80 years of age, implying that they might not need to participate at all to the breast cancer screening programme, and around 20% with the highest risk would start before 40 years of age. The principle demonstrated by Chatterjee et al. (2016) clearly illustrates that a combination of PH programmes and genetic risk stratification is conceivable.

Barriers: New Stakeholders Facing Old Questions

When PH physicians or other non-genetic healthcare workers start to work on applications of genetics in PH, they need education. Not only do they need knowledge, but also they need to know how to apply this knowledge. They may need to develop skills for family history taking and the evaluation of pedigrees, consultation skills regarding timely recognition of patients at risk and skills for organising their actual practice in such a way that their service integrates genetics. Genetic education may start from needs assessment. A Dutch project started studying needs for primary care workers, based upon which it turned out to be possible to develop effective training to identify patients at risk for hereditary cancers (Houwink et al. 2015). Curricula may focus on scientific aspects of human genetics, but priorities of non-genetic healthcare workers may be completely different, for instance knowing how to draw and discuss a family history and how to approach first-degree relatives of an affected person in an ethically sound manner. For an elaboration of training needs regarding genetics and genomics, see also Chap. 3, Sect. 3.2.

Health Technology Assessment

Now that more genes appear to be associated with breast cancer, health technology assessment (HTA) is becoming even more relevant. What are the pros and cons of specific DNA tests? Would a panel investigating several hereditary tumour syndromes (high-risk genes) be worthwhile, or are panels investigating a few specific variants that are frequent in a specific population to be preferred? Would many *variants of unknown significance* be identified in people, leading to an increase in uncertainty? Can the risk stratification be improved when family history information and PRSs based on hundreds of SNPs are integrated? What proportion of mortality in young breast cancer patients can be attributed to variants in these genes? The lack of HTA for the application of genetics is a barrier for responsible implementation (Cornel and Van El 2017; Rosso et al. 2017). For an elaboration of HTA for evaluating genetic and genomic applications in healthcare, see Chap. 3, Sect. 3.1.

Translational Research

If a health region would integrate genetic risk more into the breast cancer screening programme, this might be organised as a type of translational research. Evaluating the health outcomes would show in practice to what extent guidelines are followed, cancers are recognised at an earlier stage and at a younger age, and what extra resources are used. Where "translational research" is funded, this often refers to translation "from mice to man." A barrier of the implementation of genomic applications thus is the lack of this type of translational studies: studies where we evaluate whether we move knowledge off the shelves and into the clinic to benefit the population.

Facilitators: The Angelina Jolie Effect and Clear Evidence

Public awareness increased when the famous American actress Angelina Jolie was very open in the media about her increased risk of hereditary breast and ovarian cancers. She discussed prophylactic surgery to be able to tell her children that they need not fear that they would lose her due to breast cancer. Suddenly a scientific development reached the public debate and many women who had a family history of breast and/or ovarian cancer as well as their care providers became (more) aware of the options of DNA testing, surveillance and prophylaxis.

Furthermore, the clinical utility of BRCA testing is without doubt. The Centres for Disease Control (2014) consider HBOC due to mutations in the *BRCA1* or *BRCA2 genes* a tier 1 genomic application: "these conditions are poorly ascertained by the healthcare system, many individuals and families affected by them are not aware that they are at risk; however, early detection and intervention could significantly reduce morbidity and mortality." In terms of the innovation curve, this example moved beyond the phase of early adopters. Several primary care providers, oncologists and clinical geneticists have incorporated HBOC in their daily practice since more than a decade. They agree to do so because of the clear scientific evidence that morbidity and mortality can be reduced. Even if the integration of DNA testing for high-risk breast cancer genes in predictive genomic applications is a challenge, many colleagues in medicine already use this innovation and can help to facilitate a change in culture and structure in PH, for instance in training.

Learning from Examples

In the paragraphs above we discussed high-risk genes for breast cancer, but many similar examples exist (Rigter et al. 2014). A shift from curative to preventive services, from in-patient to out-patient treatment, from specialised genetic services to genetics as an integral part of general health services can also be pursued for cardiogenetic conditions and inherited colon cancer. Monogenic subtypes of these common disorders require a different policy, and a high-risk group can profit from a systematic approach. In the paper by Rigter et al. (2014) a cardiogenetic service in Sweden is discussed as an example, where after the sudden cardiac arrest of a young football player a multidisciplinary network was created, to recognise causes of sudden death and take care of the families.

Challenges identified in other examples of genetic services for monogenic subtypes of common disorders include "resistance to new divisions of responsibilities, and a need for more close collaboration and communication between geneticists and non-geneticists" (Rigter et al. 2014). While geneticists may have to give up some of their autonomy, PH care workers could work together with them to integrate family history taking and/or genetic testing in regular preventive services. PH genetics is an emerging field at the interface of PH and genetics, and cancer screening is one of the first activities where these fields come together. Stratified cancer screening programmes could include risk assessment based on family history, DNA testing as well as cascade testing.

6.3   Designing Systems that Integrate Genomic Information to Deliver Personalised Care

Anant Jani and Muir Gray

Health and care services exist to keep individuals and populations healthy so that they can live fulfilling lives and also contribute to building and maintaining sustainable and resilient societies.

Globally, most health and care services still make reimbursements based on fee-for-service or fee-for-activity models where service providers are held accountable for and reimbursed based on inputs and outputs rather than actual health-linked outcomes delivered to patients and populations. These models create perverse incentives and have led to several problems including overdiagnosis and overtreatment. Furthermore, these perverse incentives have meant that services are distributed and siloed, which has led to many further problems including (Jani and Gray 2015):

- Patient harm
- Unwarranted variation in outcomes
- Inequity
- Failure to prevent preventable diseases
- Waste of finite resources (financial and human)

The Role of Triple Value Healthcare

In light of these challenges, a global and EU-wide movement has been gaining momentum since 2010 to utilise value-based healthcare as a guiding principle to address the problems inherent in health and care services as they are currently organised. In the EU, value-based healthcare principles evolved in the context of universal healthcare, which has two key constraints: (1) to deliver care to the entire population and (2) to deliver this care within a finite budget. This has led to a conceptualisation of value-based healthcare in universal healthcare systems focused on three value pillars (Jani et al. 2018).

Personal Value: Value at the Level of the Patient

- Is the care that is being delivered meeting the needs and expectations of the patients to whom you are accountable?
- The needs and expectations will include objective clinical outcomes (e.g. lower blood pressures) as well as personal outcomes (e.g. being able to golf or play with one's grandchildren).

Technical Value: Value at the Level of the Intervention

- What are the patient and population level outcomes delivered by an intervention and what are the resources (money, time, space, carbon) needed to deliver those outcomes?
- In addition to examining the absolute value of an intervention, the relative value must also be considered to help in the identification and utilisation of higher value interventions as well as disinvestment from lower value interventions.

Allocative Value: Value at the Level of the Population

- What is the best way of allocating one's finite resources (money, time, space, carbon) to meet the needs of the entire population to whom you are accountable?
- A consideration also needs to be made about how to balance the individual (e.g. personal value) and population needs—everyone will not be able to get everything they want and an open and honest dialogue with the public is essential here.

Delivering Triple Value Through Healthcare Systems

When considering triple value healthcare, a key question is around the mechanisms that can be used to deliver value across the three pillars. In our work, implementation of triple value healthcare is driven through healthcare systems, which relate to networks, pathways, and healthcare services as indicated below:

- Systems: The outcomes (which are equity based and service independent) delivered to patients and populations.
- Networks: The individuals and organisations who deliver the outcomes (i.e. GPs, geneticists, surgeons, social care)
- Pathways: The means by which the outcomes are delivered.
- Healthcare Services: Evolve from the Network and Pathway structures used to deliver outcomes to patients and populations.

Through this conceptualisation of Systems-Networks-Pathways, systems can be seen as being relevant across time and space because a system focuses on the outcomes delivered to patients and populations (e.g. preventing asthma attacks, enabling people with asthma to live their life to the fullest), which means that a system will be applicable as long as the clinical condition exists—for example, people with asthma existed 50 years ago and they will exist 50 years from now and there will be people with asthma in Berlin, Munich, Toronto, London, Durban, Delhi, Sydney, etc. In contrast, the people who deliver the care (i.e. the networks) and the means they use to deliver the care (i.e. the pathways) will vary over space because of differences in demography, local politics, and resources and they will also vary over time as our understanding of the condition increases, new technologies facilitate the delivery of better outcomes and political mandates change the structure of the healthcare service. Across all of this, however, the ideal is that we adhere to considerations of health equity to ensure that a patient with a given condition should expect and receive the same outcomes irrespective of when or where they live.

Because of the focus on outcomes, systems provide a great starting point for the delivery of value but there are two further points that must be considered—the first is understanding whose perspective we are referring to when defining the outcomes and the second is to ensure that resources used to deliver the outcomes are consistently accounted for.

Defining outcomes for a system is a delicate task because of the different perspectives involved when considering a patient's or population's outcomes. We address this by working with multi-stakeholder groups that include the key stakeholders in the care pathway (i.e. patients, GPs, specialists, nurses, allied health professionals, IT, commissioners) to capture perspectives on what an ideal patient or population outcome should be and subsequently compose a shared vision on the outcomes all are aiming to achieve for that group of patients (Jani and Gray 2015). The 10-step process (Fig. 6.2) below is a method we have used in England since 2011 and as part of the PRECeDI project since 2014 to elicit and capture outcomes that matter to a multi-stakeholder group, to account for the resources being used to deliver these outcomes as well as the networks and pathways that will be used to deliver the outcomes:

Designing the system specification requires a dedicated group to take the initiative and start elaborating the subsequent steps. The first step is to define the scope of the system of care, which might be a symptom (e.g. breathlessness), a subgroup of population (e.g. frail elderly) or condition (e.g. asthma). It is also essential to define the population to be served precisely, not only by naming it but also by specifying the practice and/or local authority. From there, the multi-stakeholder group would agree the aims, objectives, outcomes-linked criteria that would facilitate measurement of progress towards the objectives, the standards linked to each outcomes-linked criterion, as well as the network, pathways and budget. A key aspect of the system specification is the requirement to produce an annual report that records data on outcomes delivered as well as resources used—thus giving an indication of the value (outcomes/resources used) of the service. As data are critical

Fig. 6.2 10-step process to design an outcomes-based system specification

for evaluating and guiding improvement, collecting specific data is critical to measure/establish the baseline, evaluate the current system, understand and identify the gaps and/or areas where a service is not doing well (e.g. underuse/underdiagnosis), and identify wasted resources. Furthermore, the data can be used to determine how a service is evolving over time and if it is actually improving. Finally, collected data can also be used to determine how a service compares to other services serving similar demographics.

Initially the data will not be perfect—it may be incomplete, and the quality may not be high. Furthermore, even when there is agreement with the objectives and criteria, getting everyone in the system to work in a coordinated way and break down artificial silos may also be difficult. However, the use of the system specification and a commitment by the stakeholders to collect and review their system's data is a critical step in starting to shift the working practices of the healthcare professionals which can lead to improved service delivery, improved outcomes and, importantly, an improved culture.

Designing Systems that Integrate Genomic Information to Deliver Personalised Care

There is much attention and excitement in the current healthcare environment on the potential of genomic information to support the delivery of PM in healthcare systems. Using our conceptualisation of healthcare systems requires that we first address whether genomic information can actually deliver value.

Personal Value: The expectations and values of the patient must be taken into account when considering the appropriateness of genomic information. For example, in some cases like Huntington's disease where the genes associated with the disease can be identified but provides no clinical utility because there is no treatment for the condition, the patient may not want to know their diagnosis because of the potential anxiety it may cause them.

Technical Value: An important consideration is the cost and benefits of genomic applications that are currently available. Costs and benefits of genomic applications can be determined through a variety of methods including using the analytic validity, clinical validity, clinical utility, and ethical, legal, and social implications (ACCE) framework, HTA (examines the properties and effects of a health technology, taking into consideration scientific, technological, medical, social, legal and ethical issues) and use of the Wilson and Junger screening criteria for the available interventions. There is no general consensus on which method is preferred when determining the value of different genomic applications, but a general approach for the system should be to decide on a payment threshold they would use to determine whether genomic information should be integrated into the system.

Allocative Value: A key constraint of universal healthcare systems is that the allocation of finite resources always comes with trade-offs and opportunity costs.

Taking this into consideration means that resources need to be allocated to deliver the best patient and population outcomes while optimising resource utilisation. It is not enough to identify the technical value of an intervention through its cost/benefit ratio because even if the intervention is of high technical value, if the wrong intervention is used, if the right interventions are not executed properly, if the wrong patient gets the intervention or if there is a higher value alternative available, the healthcare system may be wasting its resources with waste defined as: "spending on services that lack evidence of producing better health outcomes compared to less-expensive alternatives; inefficiencies in the provision of healthcare goods and services; and costs incurred while treating avoidable medical injuries, such as preventable infections in hospitals" (Lallemand 2012).

Conclusions

In this sub-chapter we have presented a conceptual framework of outcomes-based healthcare systems that can be used to deliver triple value. We have also outlined some considerations that should be made about the value of genomic information in the context of healthcare systems. We acknowledge that there is great hope and prospect for genomic information to deliver personalised care but we also highlight that we must proceed with caution before integrating genomic information into healthcare systems without understanding if this information can be used to deliver better health outcomes for patients and populations.

6.4 Clinical Decision Support Tools to Assist Physicians in the Screening of Genetic Diseases

Jim Roldan

In previous sections of this chapter the authors have discussed the importance of genetic screening to identify potentially harmful DNA mutations that can facilitate the development of certain diseases. The list of mutations that are known to be related with an increased prevalence of certain cancers and metabolic diseases is continuously growing as genetic scientists are becoming better able to identify such relations.

In the Search of Genetic-Associated Health Risks

Usually, practitioners suspect the existence of potential harmful mutations when they find an increased incidence of a particular disease within a group of family-related members. Because all of us bear two non-identical copies of genes, each one being received from one of our parents, the heritage from an individual bearing an alteration may or may not be transferred to each of the descendants with a 50% chance. So, when observing an imaginary full family tree genetic map for a particular mutation we may expect finding a mix of individuals that may carry the mutation together with other family members without the mutation.

In most cases, the presence of the mutation may not directly result in the development of the disease.

In relatively fewer cases, the affected individuals may bear not just one but two identical copies of the altered genes. This situation requires that both parents are affected by the same mutation in at least one of their DNAs and that the descendant receives the wrong copy from both parents. In this particular case the mutation may have a stronger effect on developing the disease.

For some mutations, the suspicion of the mutation can be supported by the observation of an unusual early onset of the disease when compared with the population without the particular alteration.

When observing a full cluster of family members around a particular index patient, it is usual to find patients not carrying the mutation and others carrying it. However, for most mutations it is quite possible to find patients developing the particular disease despite not having the mutation or patients having the mutation that don't develop the disease. This means that there may be other associated conditions, plus other genes influencing the risk of disease onset.

Such conditions may include the presence of other mutations that may produce a combined effect; the existence of other lifestyle or environmental factors coupling with the mutation; or even an undefined statistical increase that cannot be associated with any other particular factor, and that we regard as an increased probability chance.

The family tree screening is the usual methodology to be used to assess the risk that a potential mutation may be present in a family cluster. The generic model is based in collecting the relevant health antecedents of first-degree family members of the patient, looking for the diseases developed by each family member, the onset of such diseases and the family relation.

Assessing Potential Genetic Risk: Evidence-Based Scorings

Since hereditary diseases may generate quite different patterns in the family tree according to several factors discussed previously, some authors have tried to find a simple way to determine the likelihood of the presence of a mutation in a family cluster to differentiate it from the mere coincidence of several family members being affected by a disease for other reasons (environmental, cultural or just by chance).

The usual methodology is to collect all the occurrences of family relative's events and to weight each of the occurrences depending on the distance in the lineage, the frequency of affected members or the age of onset.

Table 6.2 Genetic screening scoring models for BRCA and familial hypercholesterolaemia

Mutation/disorder	Scorings (references)
BRCA-related cancer	Ontario Family History Assessment Tool (Gilpin et al. 2000; Nelson et al. 2013)
	Manchester Scoring System (Evans et al. 2004; Nelson et al. 2013)
	Pedigree Assessment Tool (Hoskins et al. 2006; Nelson et al. 2013)
	Family History Screen-7 (Ashton-Prolla et al. 2009; Nelson et al. 2013)
	Referral Screening Tool (Bellcross et al. 2009; Nelson et al. 2013)
Familial hypercholesterolaemia	Dutch Lipid Clinic Network criteria (Austin et al. 2004)
	Simon Broome criteria (Humphries et al. 2006)

Each proposed scoring system includes a set of thresholds that qualify a certain family cluster in different outcomes ranging from "quite probable" to "highly improbable." Those scores are usually validated with a certain population sample to determine the *sensitivity* of the scoring (the ability to detect the presence of a particular mutation) and its *specificity* (the capacity to avoid the detection of false positives—patients that have a positive score but don't have the mutation).

In Table 6.2 there is a list of different genetic scoring systems from different authors that are widely recognised for two quite well-known hereditary conditions: *BRCA*-related cancer and FH.

As it can be seen by the examples exposed, for each of the potential mutations to be screened there may be quite a few potential alternative scoring systems. Each of them having a different level of sensitivity and specificity.

Despite the fact that most of those scores may include a complex arithmetic of additions and multiplications of weights depending on different family history findings, at the end all of them use quite similar source information, such as the number of relatives affected, the age of onset of the disease and the genetic distance of the relative affected. Some scorings may take into account specific genetic ancestry (such as Ashkenazi heritage) or the severity of the symptoms (for FH).

Clinical Decision Support for Genetic Risk Scoring

The traditional manual assessment of a genetic screening assessment involves (a) collecting the family history antecedents; (b) selecting a scoring system and (c) performing the calculations to determine the scoring value and the assessed risk.

This requires a quite deep knowledge of the particular mutation and the particular scoring system selected, plus enough time to be able to perform the calculations and understand the results. This process turns to be too cumbersome for most of the primary care practitioners.

The alternative is to create a simple computer system guiding the collection of the elements that are relevant for all the assessments and allow a computer algorithm to perform all the calculations and generate a comprehensive report including all the scoring methods. This kind of computer-based support is called a *clinical decision support service* (CDSS) (Bui and Lee 1999).

The basis of CDSS is very simple: There is some interface that requests the health professional to enter the relevant information from the family tree; then the algorithm calculates a scoring following the guidelines of each of the evidence-based scoring systems available; compares the result with the thresholds set by each scoring; and returns a set of recommendations and a literature references.

As it can be seen in this process, the main purpose of the CDSS is to provide support to the practitioner, assisting him in the collection of the relevant information, performing the necessary calculations automatically and providing the advice based on evidence-based studies. In any case, the final assessment is always taken by the professional (Andrews 2013).

In order to verify the benefit of this CDSS approach, as part of the PReCEDI project we developed a CDSS for genetic disease screening that was connected to the Linkcare collaborative medicine platform. The resulting system was able to perform a complete BRCA screening in less than 30 seconds, compared with the standard 5 min that would take performing a similar assessment when performed in the manual way. Most important, the practitioners where able to execute the assessment despite not being specifically familiar with the details of BRCA screening.

From a Practitioner-Based Model to a Public Health Approach

One of the advantages of the use of CDSSs is that they can be reused by many organisations and connected to other existing PH risk assessment systems.

Instead of connecting the CDSS to a collaborative medicine tool like Linkcare, it is also possible to use it in connection with other health information systems, such as national shared medical records.

By using the historical information already stored in the public institutions to feed the CDSS, it is possible to perform a massive real-time screening of the full population and select the potential family clusters that may be affected by a particular harmful mutation at minimum cost.

References

Andrews, P. N. (2013). A discussion of clinical decision support services. *Clinical Obstetrics and Gynecology, 56*(3), 446–452.

Ashton-Prolla, P., Giacomazzi, J., Schmidt, A. V., et al. (2009). Development and validation of a simple questionnaire for the identification of hereditary breast cancer in primary care. *BMC Cancer, 9*, 283.

Austin, M. A., Hutter, C. M., Zimmern, R. L., et al. (2004). Genetic causes of monogenic heterozygous familial hypercholesterolemia: A HuGE prevalence review. *American Journal of Epidemiology, 160*, 407–420.

Bellcross, C. A., Lemke, A. A., Pape, L. S., et al. (2009). Evaluation of a breast/ovarian cancer genetics referral screening tool in a mammography population. *Genetics in Medicine, 11*, 783–789.

Bui, T., & Lee, J. (1999). An agent-based framework for building decision support systems. *Decision Support Systems, 25*(3), 225–237.

Centres for Disease Control. (2014). Tier 1 Genomics Applications and their Importance to Public Health. Retrieved from https://www.cdc.gov/genomics/implementation/toolkit/tier1.htm. Accessed 17 Sep 2019.

Chatterjee, N., Shi, J., & García-Closas, M. (2016). Developing and evaluating polygenic risk prediction models for stratified disease prevention. *Nature Reviews. Genetics, 17*, 392–406.

Collins, F. S., Green, E. D., Guttmacher, A. E., Guyer, M. S., & US National Human Genome Research Institute. (2003). A vision for the future of genomics research. *Nature, 422*, 835–847.

Cornel, M. C., & van El, C. G. (2017). Barriers and facilitating factors for implementation of genetic services: A public health perspective. *Frontiers in Public Health, 5*, 195.

Council of Europe, T. O. (2019). Additional protocol to the convention on human rights and biomedicine concerning genetic testing for health purposes. Committee on Bioethics (DH-BIO), Strasbourg, 17 May 2019. https://rm.coe.int/inf-2019-2-etat-signratif-reserves-bil-002-/16809979a8. Accessed 3 Feb 2020.

Evans, D. G., Eccles, D. M., Rahman, N., et al. (2004). A new scoring system for the chances of identifying a BRCA1/2 mutation outperforms existing models including BRCAPRO. *Journal of Medical Genetics, 41*, 474–480.

Gilpin, C. A., Carson, N., & Hunter, A. G. W. (2000). A preliminary validation of a family history assessment form to select women at risk for breast or ovarian cancer for referral to a genetics center. *Clinical Genetics, 58*(4), 299–308.

Hoskins, K. F., Zwaagstra, A., & Ranz, M. (2006). Validation of a tool for identifying women at high risk for hereditary breast cancer in population-based screening. *Cancer, 107*, 1769–1776.

Houwink, E. J., Muijtjens, A. M., van Teeffelen, S. R., et al. (2015). Effect of comprehensive oncogenetics training interventions for general practitioners, evaluated at multiple performance levels. *PLoS One, 10*, e0122648.

Humphries, S. E., Whittall, R. A., Hubbart, C. S., et al. (2006). Simon Broome Familial Hyperlipidaemia Register Group and Scientific Steering Committee Genetic causes of familial hypercholesterolaemia in patients in the UK: Relation to plasma lipid levels and coronary heart disease risk. *Journal of Medical Genetics, 43*, 943–949.

Jani, A., & Gray, M. (2015). Outcomes as a foundation for designing and building population healthcare systems in England. *BMJ Outcomes*, 16–19.

Jani, A., Jungmann, S., & Gray, M. (2018). Shifting to triple value healthcare: Reflections from England. *Zeitschrift für Evidenz, Fortbildung und Qualität im Gesundheitswesen, 130*, 2–7.

Khoury, M. J., Gwinn, M., Yoon, P. W., et al. (2007). The continuum of translation research in genomic medicine: How can we accelerate the appropriate integration of human genome discoveries into health care and disease prevention? *Genetics in Medicine, 9*, 665–674.

La Duca H, Polley EC, Yussuf A, *et al.* (2020) A clinical guide to hereditary cancer panel testing: Evaluation of gene-specific cancer associations and sensitivity of genetic testing criteria in a cohort of 165,000 high-risk patients. Genetics in Medicine. 2020;22(2):407-415. doi: https://doi.org/10.1038/s41436-019-0633-8

Lallemand NC. (2012). Reducing waste in health care. *Health Affair*. Retrieved from https://www.healthaffairs.org/do/10.1377/hpb20121213.959735/full/. Accessed 23 Sep 2019.

Little, J., Potter, B., Allanson, J., et al. (2009). Canada: Public health genomics. *Public Health Genomics, 12*(2), 112–120.

Mavaddat, N., Michailidou, K., Dennis, J., et al. (2019). Polygenic risk scores for prediction of breast cancer and breast cancer subtypes. *American Journal of Human Genetics, 104*(1), 21–34.

Nelson, H. D., Fu, R., Goddard, K., et al. (2013). *Risk assessment, genetic counseling, and genetic testing for BRCA-related cancer: Systematic review to update the U.S. preventive services task force recommendation [internet].* Rockville, MD: Agency for Healthcare Research and Quality (US); (Evidence Syntheses, No. 101.).

Phillips, A. M. (2016). Only a click away - DTC genetics for ancestry, health, love…and more: A view of the business and regulatory landscape. *Applied & Translational Genomics, 8,* 16–22.

Pohlhaus, J. R., & Cook-Deegan, R. M. (2008). Genomics research: World survey of public funding. *BMC Genomics, 9,* 472.

Rigter, T., Henneman, L., Broerse, J. E., et al. (2014). Developing a framework for implementation of genetic services: Learning from examples of testing for monogenic forms of common diseases. *Journal of Community Genetics, 5,* 337–347.

Rosso, A., Pitini, E., D'Andrea, E., et al. (2017). The cost-effectiveness of genetic screening for familial hypercholesterolemia: A systematic review. *Annali di Igiene, 29,* 464–480.

Shah, T. A., & Guraya, S. S. (2017). Breast cancer screening programs: Review of merits, demerits, and recent recommendations practiced across the world. *Journal of Microscopy and Ultrastructure, 5,* 59–69.

Unim, B., Lagerberg, T., Pitini, E., et al. (2017). Identification of delivery models for the provision of predictive genetic testing in Europe: Protocol for a multicentre qualitative study and a systematic review of the literature. *Frontiers in Public Health, 5,* 223.

Unim, B., Pitini, E., Lagerberg, T., et al. (2019). Current genetic service delivery models for the provision of genetic testing in Europe: A systematic review of the literature. *Frontiers in Genetics, 10,* 552.

Washington State Department of Health, Genetic Services Section; Health Resources and Services Administration, Maternal and Child Health Bureau, Genetic Services Branch. (2008). Final Report of Genetic Services Policy Project (GSPP). Retrieved from www.mchlibrary.org/collections/docs/U35MC02601.pdf. Accessed 30 Dec 2019.

Index

Printed in the United States
By Bookmasters